Contents

Agricultural Revolution in Africa ?

Impediments to change and implications for farming, for education and for society

R.G. Lowe

MACMILLAN
PUBLISHERS

First published 1986

Published by *Macmillian Publishers Ltd*
London and Basingstoke
Associated companies and representatives in Accra, Auckland, Delhi, Dublin, Gaborone, Hamburg, Harare, Hong Kong, Kuala Lumpur, Lagos, Manzini, Melbourne, Mexico City, Nairobi, New York, Singapore, Tokyo

ISBN 0-333-42455-7

Printed in Hong Kong

British Library Cataloguing in Publication Data
Lowe, R.G.
Agricultural revolution in Africa?:
impediments to change and implications
for farming, for education and for society.
1. Agricultural productivity—Africa
I. Title
338.1′6′096 HD2117

ISBN 0–333–42455–7

Dedication

To the memory of my mother,
who made me what I am.

Preface

This book has evolved from a series of lectures on multiple land-use given at the University of Ibadan. The lecturer may learn more from giving his lectures than his students do from hearing them – because he is compelled to consider carefully why he thinks what he thinks. It was necessary to ask three questions:

- why is freehold tenure widely believed to be more conducive to land development than communal tenure?
- why is owner-cultivation often regarded as the most efficient system of farming?
- what contribution do cooperatives make to progress in agriculture?

and to look at the constraints to implementing these concepts in tropical Africa. Also, in seeking the answers, it was clearly necessary to look at the European experience, which influences so much of our present thinking, and to ask the extent to which this is applicable in Africa.

Most of my working life has been spent in West Africa, and much of it in various parts of Nigeria. Although emphasis is laid on personal experience, I have called on other authors to strengthen the argument. Nigeria will, perhaps, be the first country in West Africa to experience the trauma of agrarian revolution, as peasant farming gives way to commercial agriculture. Nigeria has the largest population and the largest economy in 'Black Africa'. The average population density is already comparable to that of Europe: equal to that of France, almost half that of Great Britain today, and equal to that of Great Britain in the mid-nineteenth century. Moreover Nigeria is an extensive country, and contains within its borders a considerable range of the land types that typify tropical Africa, from arid savanna to equatorial rain forest. It is likely to establish the pattern of development for the rest of West Africa; and there are lessons to be learned by other tropical African countries, and perhaps by developing countries in other parts of the world. Nor are the difficulties of Nigeria without significance for the industrial countries.

In this book, the nature of agrarian revolution is discussed, together with its repercussions for farming, for education, and for society.

R. G. Lowe

Department of Forest Resources Management,
University of Ibadan

Acknowledgements

I should like to express my gratitude to Professor S.K. Adeyoju and Professor D.U.U. Okali, of the Department of Forest Resources Management at the University of Ibadan, for the encouragement which eventually led to this book being written; to Dr H.J.W. Mutsaers of the International Institute of Tropical Agriculture, and Professor S.K. Adeyoju for comments on the original draft; and to my wife for much patience.

Photographic acknowledgements

The author and publishers wish to acknowledge, with thanks, the following photographic sources.
J. Allan Cash; FAO (photographs by Banoun, F. Botts, A. Defever, E. Syts); Firestone Rubber Company; Professor R.J. Harrison Church; R.G. Lowe; World Bank.
The publishers have made every effort to trace the copyright holders, but if they have inadvertently overlooked any, they will be pleased to make the necessary arrangements at the first opportunity.

1 The problem

'Of course the era of the hoe will very soon come to an end, but it would be wise if the agronomists could keep this humble tool before their eyes as a reminder of reality.'

Sylvia Leith-Ross

In the oil-fuelled economy of Nigeria peasants have been deserting their farmlands in ever greater numbers. In 1960, about three-quarters of the working population was engaged in agriculture; in 1985 the figure may have fallen to less than half, although severe economic recession is making alternative employment difficult to find. Even in Ghana, which has no recourse to petrodollars, the number of farmers has declined with disastrous consequences for the national economy (see *Table 1*). Nevertheless, in many countries, of both West and East Africa, food production is insufficient to meet the needs of their populations. In Nigeria it is estimated that in 1976 the country imported 6.8 per cent of its food supplies in calorific terms, and that in 1985 this could reach 24 per cent of gross supply.[1]

The principal source of these difficulties is biological, attributable to the soil characteristics of the humid tropics. Peasant agriculture was adapted to these constraints, but shifting cultivation and related fallowing systems have recently been described by Professor B.N. Okigbo, Deputy Director General of the International Institute of Tropical Agriculture (IITA) at Ibadan, as 'outmoded farming systems'. Professor Okigbo emphasises that these farming systems are ecologically sound for subsistence agriculture under low population densities and long-term fallows, but under various socio-economic pressures they are now breaking down. For example in Nigeria the

1 Idachaba, F.S. & others (1980) *The Green Revolution: a food production plan for Nigeria (Final Report)*, Federal Ministry of Agriculture, Lagos

Table 1 Decline of cocoa production in Ghana since 1970 (thousands of tonnes)

Year	Cameroon	Ghana	Ivory Coast
1982/83 (forecast)	112	183	380
1981/82 (estimate)	122	225	457
1980/81	120	258	403
1979/80	124	285	379
1978/79	106	250	312
1977/78	107	268	304
1976/77	82	320	230
1975/76	96	397	231
1974/75	118	377	242
1973/74	110	350	209
1972/73	107	418	181
1971/72	123	464	226
1970/71	112	392	180
Gill & Duffas Cocoa Statistics, May 1983			

population in 1900 was only about 10 million, whereas in 1985 it is estimated to be almost 100 million. The total land area of Nigeria is 92.3 million hectares, of which about half may be available for arable cultivation. If this land were cultivated for two years and then cultivation were abandoned for say six years, as under traditional fallowing systems, then less than a fifth of a hectare per head would be available for arable cultivation. In some densely populated areas the resting period for land has dwindled to only two years. Short-term fallows and extended periods of cultivation are resulting in erosion, soil degradation and loss of productivity.

In Nigeria, peasants still produce 97 per cent of the total domestic food production, and this is the general pattern throughout West Africa. Peasant farms are small, and traditionally are entirely cultivated with hand tools – the axe, the short-handled hoe and the matchet (or cutlass). In forest areas, a man with one wife can farm between 0.2 and one hectare of new land each year, and in savanna areas perhaps up to two hectares, while maintaining the farm from the previous year or two years. Where a peasant has tree crops, such as cocoa or rubber, these average about two hectares, though they may extend to as many as twelve or more hectares.

In some countries of East and Central Africa, such as Kenya or Zimbabwe, a substantial proportion of food production is due to large-scale commercial farming, initiated by Europeans. In these countries, the conditions are perhaps more favourable to mechanised agriculture, associated with drier climate, or high altitude, or volcanic soils, and a relatively sparse population. The use of draught animals or tractors can increase the area cultivated. In some savanna localities in West Africa, particularly on loess soils, during the past three decades the ox plough has been successfully introduced, and one team can cultivate up to about 15 hectares. This has been most successful where the inhabitants have a tradition of both cattle rearing and arable cultivation, as for example amongst the settled Fulani in the Gombe District of Nigeria. Using such methods, small-scale farmers in Zimbabwe have recently been able to achieve spectacular improvements in production, associated with better cultural techniques, better farm inputs such as fertilisers, and better access to credit. Horses can work longer hours than oxen, but require better nutrition due to their less efficient digestive systems; they cannot be kept fit for the effort of ploughing on the poor feed available to peasant farmers, particularly before the onset of the rainy season when fodder is scarce. Moreover, horses are more sensitive to Trypanosomiasis (sleeping sickness), and are not used for farm traction in lowland tropical Africa. In forest regions, cattle are not used as draught animals, because of the greater prevalence in this zone of Tsetse fly, the carrier of Trypanosomiasis. Peasant farms are too small to support a tractor, and light mechanical cultivators may be unable to cope with compact soils, or with tree stumps and roots remaining in the ground – not to speak of the difficulties for peasants to maintain mechanical equipment in rural areas. Government tractor hire units have not been a success, due to the problems of government enterprise discussed in later chapters of this book; in Nigeria they are being phased out.

The increasing area of cultivated land reduces the land available for free-ranging livestock. These may either be owned by nomadic herdsmen, such as Fulani and Arab descended peoples who do not themselves own land, or they may belong to peasant villagers. In forest regions the farms tend to be more than one kilometre from the village, to protect unfenced fields from wandering sheep and goats, but are not usually more than five kilometres away – about an hour's walk. Reduction in the grazing area eventually causes overgrazing, and deterioration of the pasture, with a reduction in the numbers and quality of the stock that can be maintained, particularly through the dry season. In more densely populated localities, it has become

Peasant farmers hoe their fields

Ploughing with a tractor

Fulani herdsman with his cattle

necessary to confine cattle or goats during the growing season, to prevent damage to crops, and to practise stall feeding. However there is an ever-enlarging demand for meat and animal products by burgeoning urban communities.

A report states, 'The production of milk is given high priority by many concerned with human nutrition in Nigeria. It is essential to recognise that, under the present conditions ... the production of marketed milk from the traditional sector is incompatible with the demand for improved production of weaners and slaughter cattle ... In short, increased production of milk for sale can only be achieved by depriving calves of food essential for proper growth...'. With regard to beef production it continues, 'In contrast to the apparently low cost of production in the traditional sector, the cost of operating these [cattle] ranches has been high ... The economic viability of ranches in Nigeria and elsewhere in Africa depends on a reliable supply of relatively cheap immature cattle, cheap supplementary feed and on a sufficiently high price for meat.' The report points out that pigs and poultry (unlike cattle, which are ruminants) cannot utilise non-protein nitrogen or digest bulky cellulose-rich material and are vulnerable to shortages in the supply and price of grain. The report recommends that all schemes to promote milk collection from traditionally managed herds should be abandoned, and that proposals for establishing dairy farms or conventional fattening ranches should be re-examined; and, in concluding, the author of the report feels obliged to state, 'It is with dismay that the writer realises that only after twenty-five years has he begun to understand the constraints to animal production . . .'.

Tropical agriculturists may be surprised by the importance of forestry in the industrialised countries. The proportion of total land area covered by forest is, for example, France 25%, West Germany 29%, United States of America 33% and Japan 67%. In the United Kingdom only 9% of the land area is dedicated to forestry, and the country imports 90% of its wood requirements. In Nigeria in 1919, a target of 25% of the land area was to be set aside as forest reserves.[2] In fact only 10% has been reserved, and FAO have recently estimated that before the year 2000 Nigeria will have exhausted its existing high forest resources.[3] Savanna woodland covers four-fifths of the total reserved area, and the demand for fuelwood is sufficient to consume the whole of this in less than five years.

Even 40 years after exploitation, savanna woodland does not recover its original density. The FAO report states, 'Since the natural forest within forest reserves is insufficient to meet the country's requirements and since the forest areas outside reserves are a fast diminishing asset, there is need to expand the resource base.' The report recommends an attempt to create more forest reserves, that de-reservation of existing areas should be resisted, and that encouragement should be given to private and commercial enterprise to establish timber crop plantations, presumably on private land. Nevertheless wasteful farming increases the pressure to de-reserve forest reserves. There has also been a tendency for Government to appropriate parts of forest reserves for other forms of land use, such as grazing, farm settlements, or plantation crops such as oil palm or rubber. This may be for no better reasons than that they are consolidated blocks of land over which Government exerts some control, and that it evades the troublesome problem of land tenure. Yet development programmes have been less concerned with increasing yields of existing farms than with extending cultivated areas.

The pressure on land is distributed very unequally over the country

2 Lugard, F.D. (1919) *Political memoranda, No.XIII – Forestry*, Waterlow, reprinted by Cass, 1970

3 FAO (1979) *Forestry development Nigeria: project findings and recommendations*, Rome (p. 88)

Table 2 Nigeria: population, area and land/man ratio, by States

States	**Percentage of total population**	**Area (km²)**	**Estimated population (millions)**	**Land area per capita (hectares)**
Northern States				
Sokoto	7.45	102 620	7.150	1.43
Kaduna	7.20	70 300	6.900	1.02
Kano	10.15	43 320	9.750	0.44
Bauchi	4.30	64 660	4.100	1.57
Borno	5.25	116 500	5.050	2.30
Central States				
Kwara	3.00	66 930	2.900	2.31
Niger	2.10	65 090	2.000	3.23
Abuja (FCT)	.50	6 540	0.500	1.33
Plateau	3.55	58 080	3.400	1.70
Benue	4.25	45 210	4.100	1.10
Gongola	4.60	91 470	4.400	2.08
Southern States				
Oyo	9.15	37 740	8.800	.43
Ogun	2.75	16 770	2.600	.64
Lagos	4.70	3 350	4.500	.07
Ondo	4.80	20 980	4.600	.45
Bendel	4.30	35 530	4.150	.85
Rivers	3.00	21 870	2.900	.75
Anambra	6.35	17 690	6.100	.29
Imo	6.45	11 860	6.200	.19
Cross River	6.15	27 260	5.900	.46
Total	100.00	923 770	96.000	(0.96)

Note: Within each group, States are listed approximately from west to east.

(see *Table 2*). In some regions there is a desperate shortage of farmland whereas in others, particularly in southern guinea savannas, extensive tracts are more or less devoid of population. The situation is made worse by the migration of people from less populated to more populated parts, for example towards the southern coastal towns of West Africa. Communal land tenure, the difficulties of transferring title to individuals or between individuals, and the reluctance of native communities to lose control over land which they regard as traditionally theirs, are major factors inhibiting capital investment in agricultural development. It may be difficult or tedious to obtain undisputed title to land, or to show satisfactory proof of title. It is also difficult for people who belong to one community to farm or develop lands that belong to other communities. These factors may also give land an exaggerated value. The rights to cultivate the land for arable crops and the right to usufruct of trees growing on the land may belong to different people, or may vary depending on whether a piece of land is being cultivated or lying fallow. An individual may be unable to sell, or otherwise alienate his rights in land, without the permission of his community. This web of rights in land hinders capital investment, and prevents land passing to its most economic use, depending on soil, climate and local market forces.

In a traditional society, farming system and social system are meshed together. One cannot be changed without disrupting the other. Traditional farming does respond to economic forces, but only to the extent that the whole socio-agronomic structure is not overthrown. In Africa, both the methods of farming and society itself are under stress. Peasant agriculture diminishes, rather than enhances, the fertility of the soil, and higher population densities prevent sufficiently long fallows to allow soil recovery. Moreover peasant farming does not pay well enough to satisfy aspirations for better living standards. People, and especially young people, wish to escape the stultifying bonds of traditional society, and to have a chance of enjoying the material benefits of urban industrial society. In fact, farming can only become adequately profitable when fewer farmers feed more people who are engaged in other productive activities. Industrialisation and modernisation of agriculture go hand in hand.

Methods which have been tried in Nigeria to boost agricultural productivity bring to mind observations on American agriculture, where land was appropriated by European pioneers, and the situation was very different from that in Nigeria. Discussing agricultural development in the United States during the nineteenth century, Galbraith writes that it called for 'a land policy which would get the land settled and ploughed and a transportation system which would get the products to market'. He continues, 'It was our unquestioned good fortune that community education experts, grain marketing analysts, home economists, vocational counsellors, communications specialists or public safety advisers had not been invented. Had these existed, attention would have been drawn from the strategically central tasks of getting farms settled and the railways built. And they would have been a burden on the backs of people who could not afford such luxuries.'[4] Warriner writes, 'the root of the trouble in the conception [of integral reform] lies in the attempt to transplant institutions which have been evolved in a country at an advanced stage of development to countries at more backward stages, in the belief that these institutions have been the causes of America's rapid growth and not, as they in fact are, the outcome of that growth ... Integral reform is really a social welfare policy, designed to help poor families; and its advocates neglect the economic base because they do not expect that the recipients of aid will ever become independent farm producers. This is a perfectly legitimate approach in advanced countries with localised agrarian poverty.'[5]

Writing in 1958, on his experience at the University (College) of Ibadan, Mellanby observed, 'A college faculty of agriculture in Nigeria has ... a different problem from that which faces a similar faculty in Britain or America. There is in Britain a considerable number of students who are farmers' sons who wish to return to their farms with better and more up-to-date knowledge to enable them to farm their own land more profitably. No agricultural student in Ibadan is, or is likely to be in the near future, a farmer, or has any intention of going in for farming himself. (Even if he had it at the start of his course, he would be unlikely to be confirmed in his decision.) Almost the only outlet for agricultural graduates is in the government advisory and research services. These are in fact more useful than I may have implied, but they still cannot always give really useful advice. Agricultural education at the university level cannot yet lead a "back-to-the-land" campaign.' He also writes, 'I remember Dr Stopford ... who was formerly Principal of Achimota College in the Gold Coast [Ghana] saying how irritated he was by European officials criticising his pupils for not taking up farming. He asked the government Department of Agriculture to tell him what sort and what size of farm a young man should aim at having to enjoy a standard of living similar to that of a salaried official. No-one could give him the answer.'[6]

On the subject of the university farm, Mellanby says, 'A great deal of energy was devoted to developing the farm so urgently needed for

4 Galbraith, J.K. (1964) *Economic development*, Harvard University Press
5 Warriner, D. (1969) *Land reform in principle and practice*, Oxford
6 Mellanby, K. (1958) *The birth of Nigeria's university*, Chapter 10, *Agriculture* (pp. 191–203), Ibadan University Press

teaching and research. It was this farm which served to impress on me how little real knowledge existed about the fundamental problems of agriculture in West Africa. In the first instance it was planned by people with a good deal of experience of local conditions, and their idea was to have a large demonstration farm which could show the students improved methods carried out on a large scale. It was believed that by so doing the farm would in addition prove itself an economic asset, and produce crops which would supply the college with fresh food. We even hoped that the cash value of the crops would cover a very substantial part of the cost of running the farm. The results were quite different. At no time was the value of the produce of the farm more than a small fraction of its cost of production, making no allowance for the value of the land, for the capital equipment available or for the salaries of the senior staff concerned. Also the improved methods used did not by any means always produce better crops than the local "primitive" methods used by the peasants in nearby areas. The farm always *looked* much more efficient, but the yields per acre were by no means always higher. This disappointing result was not the result of inefficiency on the part of those planning and managing the farm but simply that the knowledge of how this type of farming can be made an economic success is not available. The surprising thing is that anyone with any experience should have thought that it did. If the farm was more of a liability than we anticipated, it did at least underline the need for fundamental research in crop production.'

It must be confessed that, more than twenty-five years later, the problems remain much the same. Nevertheless there has been significant improvement in scientific and technical knowledge during the intervening period. Moreover, the economic and social situations are also transformed, particularly in Nigeria as a result of the oil boom years. Perhaps the most interesting developments on the university farm were in animal husbandry. A dairy herd was built up, despite Trypanosomiasis, and also an extensive poultry farm from British day-old chicks. The chickens did well under Nigerian conditions, and the pigs also flourished. Mellanby continues, 'I think that experimental work with animals showed that they might be kept economically at Ibadan. Of course on an experimental farm where records must be kept scientifically and where many extra staff are required, costs must be high, but the margin between the value of the produce and the costs of producing it were generally much lower for poultry and cattle than in the case of maize or other (arable) crops.'

Mellanby comments, 'One reason why costs on our farm were too high was that we tended to follow the usual custom of the country and employ far too many people for any one job . . . This sort of thing is difficult to stop. Even on the animal husbandry side of our establishment we tended to employ a dozen able-bodied men each to do one particular thing, whereas in England a single land girl could have done the lot more efficiently. For arable farming we had an excellent supply of machinery, and this was on the whole used efficiently but we also had a labour force sufficient for a much larger quite unmechanised farm.' He also writes, 'The main reason therefore for the failure in Nigeria of "up-to-date" farming is high production costs together with low yields. Nigeria is a country of peasant subsistence farmers with very small cash incomes. Farming has consisted of clearing an area of bush, hoeing the ground, planting a crop, an occasional weeding, and harvesting. This has meant, at clearing and planting, some days of hard labour and the weeding and harvesting has been moderately strenuous, but throughout the year only a small fraction of the time has been devoted to work on the farm. Nevertheless the farmer is always available when work has to be done. He can take immediate advantage of an early rainstorm, and have his planting done at the most advantageous time. He is personally interested, and sees that the weeding is done as soon as it is necessary. He is there to harvest the crop at the right time. Working with paid labour is very different. Either one has a permanent labour force which is sufficient to do the work at these times when there is much to be done and which is uneconomically employed for the greater part of the year, or else one has too few workers just when there is a rush period. Reliance has to be placed on casual labour which is devoid of the essential skill for the job and is not interested in the result.'

Sylvia Leith-Ross, discussing the Nigerian Press in 1960, wrote, 'Some of the articles and letters were very good: well-expressed, to the point, and full of sound good sense, though it was often "they" who should do this or that, seldom the writer himself. There was an article by a Lecturer in Agronomy which contained the much-needed warning: "Development and expansion of cash crops and industries must not be allowed to prejudice food supplies." He advocated, most rightly, more cooperatives, mechanisation ("else we starve"), the provision by Government of research workers and laboratories. It was all perfectly sensible and yet one felt that the writer, like so many others, thought in terms of the superstructure. No one bent down and picked up a hoe. I remember Bishop Onyeabo at Onitsha, already an old man, deploring the drift from the farms, his own arms full of the yams he had himself grown as an example to his people. Of course the era of the hoe will very soon come to an end, but it would be wise if the agronomists could keep this humble tool before their eyes as a reminder of reality. Else "Agronomy", with its flavour of textbooks and classrooms, will become like "Hygiene", a word on a blackboard, unrelated to living.'[7]

In 1980 Idachaba comments on large scale mechanised government farms, established in Nigeria during the 1976–1980 plan period, that though they were highly visible and politically appealing, they were not a success. Government bureaucracy was unsuited to management of food production and marketing. The overvalued (Naira) currency made machines relatively cheap, but heavy machinery damaged the soil, made a heavy demand on support services for maintenance and repair, and the yields realised on some of these farms were less than the national average. The farms might look technically efficient, but were not necessarily economically efficient, and they resulted in substantial social dislocations because of large scale displacements of human settlements.[8]

Private investment in extensive arable crop agriculture has not been forthcoming on any considerable scale. Permanent cultivation in the lowland humid tropics is of a greater order of biological difficulty than in temperate regions, and there is therefore a greater risk of business failure. Other reasons are the difficulties of obtaining satisfactory title to tracts of land, the low rates of return on capital invested compared with commerce and industry, and perhaps because satisfactory systems of farming have not yet been sufficiently worked out.

The African concept of farming has to be transformed. Improvement in productivity depends on developing farming systems that will increase the biological productivity of the land, rather than reducing it as does peasant farming. Then more can be siphoned off for human consumption. However, in tropical Africa, farming systems must be robust – so that when the fertilisers do not arrive, or spare parts are not available for the tractor, or foodstuffs for livestock cannot be bought, these are not an irredeemable disaster. Familiar crops, familiar livestock and familiar skills of the locality should be used to the full, and as far as possible the farmer should retain his self-sufficiency. The ownership of land should be modified so that it can finance its own development. At present, technical aid projects in agriculture tend to be highly dependent on inputs and infrastructural support that cannot be sustained, and to look to government rather

7 Leith-Ross, Sylvia (1983) *Stepping-stones – Memoirs of Colonial Nigeria 1907–1960*, Peter Owen, London (p. 189)
8 Idachaba, F.S. & others (1980) *op. cit.*

than to individuals for innovation and support. Perhaps their ideal would be groups of illiterate or semi-educated farmers who follow faithfully the guidance given to them by expert counsellors provided by government! But developed farming requires an educated rural population, who will want sufficient earnings to maintain a satisfactory standard of living. The economists and planners, who will approve the expenditure in some distant city, want projects that look tidy and convincing, and that might satisfy an accountant or banker. But nature red in tooth and claw will be the final arbiter. Douglas Hague writes, 'The reason why economists got it wrong then [in the 1930s and 1940s] was that they largely ignored [progress in] science and technology. That is why most economists still get it wrong today.'[9] Fortunately there are West Africans, and Nigerians are among them, who are amongst the most enterprising people in the world. Once having seen the light, they will not be slow to take advantage of it – provided they are not frustrated by maladroit government.

One certain thing is that agriculture is in a state of change, whether it is wanted or not. As everywhere, the need is to put men, land, capital and 'know-how' together in the right proportions. At present the main deficiency is know-how, and Africans have no alternative but to try to develop it for themselves. The problem with traditional farming is that its flexibility is too limited; but there is no virtue in replacing it with systems that will themselves prove restrictive, and unable to evolve sufficiently. Africa is launched on its 'Agrarian revolution'. This is not a result of national policy or planning; it is in the natural course of events. Governments are like midwives: they may ease the delivery or make it more painful. If, from preconceived notions, they stifle human inventiveness and persistence, it is more likely to be painful – but it cannot be halted.

9 Hague, Douglas, quoted in *New Scientist* No 1407 of 26 April 1984, pages 24–25, from Mond lecture at University of Manchester

2 The social impediment

'So long as agriculture in Nigeria and other parts of West Africa continues to offer no other equipment than a cutlass, a hoe and an axe for no reward than merely to keep body and soul together, so long will it continue to offer no attraction to the primary or medium school-leaver, or even to an illiterate young African as a means of livelihood; and so long will the army of unemployed continue to mount in the towns and cities.'

Victor Oyenuga

Peasant farming is more than a livelihood: it is a way of life. Land is regarded as something that supports a community, rather than as a source of profit. In all traditional societies, farming systems and social systems are closely intermeshed. Though African tribes differ widely from one another in their original cultural and political systems – even those occupying adjacent territories – and generalising is not easy, nevertheless where the farming system is tied to a common environment there tend to be common patterns of social behaviour. African peasants feel a mystical need to be in harmony with their environment, stemming from traditional religious beliefs. A stranger – an agricultural extension worker or a visiting expert – may disturb this well-being with calamitous consequences: a plague of locusts, for example, or destruction of a crop by rodents. Although he might bring good, he can be at the mercy of chance events, which may be wrongly attributed to his interference. In modern times these beliefs lose strength, but they may still underlie human behaviour.

In pre-Keynesian societies, wealth is regarded as finite: if one individual has more then others must have less, and society attempts to protect itself against this. Among the Etun in Cameroon, the concept could even be extended to the use of artificial fertilisers, which could be perceived as a way of stealing fertility from other people's fields. In rural Africa, social systems were geared to disperse wealth and to prevent an individual enriching himself at the expense of the community. The fortune of an individual depended on his successfully propitiating the beneficent forces and placating the malevolent forces that beset him. Clearly a person who could grow in power and wealth, when others were suffering and impoverished, was exploiting the dark powers and must be guilty of wickedness.

An individual attempting to improve himself might find sanctions taken against him. These could be physical or occult: his livestock might be poisoned or his crops spoiled; or he could suspect that misfortune in his family was wished on him by the sorcery of jealous neighbours. He himself could be accused of witchcraft or other misdeed and a case brought against him, which would at best involve him in expense or at worst place him in jeopardy. In African justice, both the innocent and the guilty are taken to task, and may suffer a public review of their characters. Although the latter will usually be compelled to make restitution to the former, both will be expected to make gifts to the court. This helps to reinforce social control, and may induce individuals to appease their fellows.

African customary law

In traditional African society, disputes between members of a descent group are taken to the head and elders of the group. When the elders fail to obtain an agreement, they take the case to the chief in charge of

A chief in council

the ward, and if he is unable to settle it he refers it to the highest judicial council of the town. In most cases this was the king sitting in council with his senior chiefs. In illiterate societies, customary law was neither codified nor based on a case law of precedents. Judges did not refer to previous judgements, though they would be conscious of their own judgements in previous years and endeavour to be consistent. Before announcing their findings, they would discuss analagous cases which came to their minds, but these would not be cited as a reason for their decision in the case before them. The objective of the law is to obtain conciliation between individuals. A court action over land may develop out of a dispute over an entirely different matter, such as a conflict over a woman, and though the details may not be mentioned in court they will probably be known to the judges. A litigant whose general behaviour in the community is thought to be reprehensible, may receive punishment out of proportion to the offence that is being heard by the court. Lloyd writes, 'In many ways customary law is more adaptable in a rapidly changing society than is codified or case law, for the judges merely forget past judgements which seem unsuited to modern problems; they are not saddled with the task of formulating qualifying arguments or devising new arguments which will restore justice to an archaic law. Customary law needs no lawyers since all know the law; and since it has no lawyers it does not develop into an esoteric science.' Lloyd also observes, 'One cannot radically change social structure by legislation, for people will ignore laws which conflict with the norms expressed by society; new laws in fact usually reflect the changes in social structure which have already taken place, by giving legal sanction to the norms which have recently become accepted by a majority of people.'[1]

Fadipe writes, '. . . in pre-colonial days . . . the powerful and often their near relatives as well had only their consciences to stay their hands in the commission of acts of injustice. They were not always subject to the regular operation of the law. If they chose to defy the law there was no immediate machinery for bringing them to book by the customary process. Nemesis overtook them only when they fell foul of some person of their own rank who not only retaliated but, if determined enough, rallied the discontented masses around him for revenge.'[2] He also writes, 'The acceptance of gifts of money and other commodities from parties in trouble with a view to obtaining leniency was not considered reprehensible. Rather it was looked upon by the

1 Lloyd, P.C. (1962) *Yoruba land law*, Oxford University Press (p. 378)
2 Fadipe, N.A. (1970) *The sociology of the Yoruba*, Ibadan University Press

authorities as a legitimate source of revenue. The unfortunate party was bled right and left by the chiefs. After he had satisfied one, another was pointed out to him as worth cultivating ... However, these gifts to members of a judicial tribunal very rarely led to the guilty party being adjudged the winner in a case of a dispute between two parties.'

In some regions, the traditional forms of African customary law have been replaced by Islamic law, which is a written codified law that guides the *alkali* (or magistrate). The change often occurred relatively recently, that is post 1800. However, appeals may go to the city ruler and his council of chiefs, who may judge the case according to equity. The best Moslem judges are of high integrity and actuated by an enlightened tradition. 'Justice is of God, and no man should derive profit from dispensing it.'[3] It may be of interest that, in Nigeria, the burgeoning of dishonesty during the past decade has less influenced the judiciary than it has many other institutions. This doubtless reflects the serious regard that Africans traditionally have for justice, and perhaps particularly by those who have it as their trust. Modern African States may retain a tendency to revert to tribunals, and to dispense with lawyers and with legal systems based on European procedures. This may be a consequence of the traditional African conception of the law as being of the people, rather than for the people.

African chieftaincy

A chief was not a capitalist autocrat, but a focal point of wealth to be redistributed among his people; his powers were circumscribed by ritual and tradition. For example, among the Yoruba of Nigeria, the liberalities expected of a chief during his lifetime, the demands of his many wives and children, the final bankrupting expense of his funeral and the rotation of the chieftaincy to another family meant that little advantage passed to his heirs. Formerly, none of the wealth of a Yoruba king could be shared among his children, but passed entirely to his successor to the throne. Lloyd points out that there were definite disadvantages in being acknowledged as belonging to the royal house, and therefore enjoying the possibility of becoming elected king. Princes who competed unsuccessfully for the throne and refused to accept the new *oba* (king) were driven with their supporters from the

3 Quoted by Lugard, F.D. (1922) *The Dual Mandate in British Tropical Africa,* reprinted by Cass, 1965

kingdom, and were received by the rulers of neighbouring kingdoms as commoners. Important but less rebellious princes were sent to the margins of the kingdom to govern an existing settlement, or with their followers to create a new one. An individual of the royal line, but with little chance of becoming king, obtained useful rights over land, but was debarred from important political office and thus gained no share of the wealth accruing to the throne. However, after two or three generations, he would be absorbed into his mother's group and no longer be regarded as a member of the royal descent. The consensus of Yoruba opinion is that the *oba* 'belongs to the people' and that therefore the palace belongs to them too. Any abnormal use of palace land must be sanctioned by the chiefs, and any profits accruing to the throne belong to the community and not to the *oba* in his private capacity – and still less to the royal descent group. There is a delicate balance between the power of the chiefs, who make the policy, and the *oba* whose sacred status commands the authority that will ensure obedience. If the chiefs felt that the *oba* was misusing his constitutional powers or governing badly, they could refuse to cooperate with him or, as the ultimate sanction, 'ask him to die', and if he refused, murder him.

Among many African peoples the king led a cloistered life. For example among the Igarra, the Attah of Idah was supposed to touch earth and water (with his feet) only once, at his formal installation.[4] The seclusion of kings was not uncommon among African peoples, and they emerged from their palaces only for specific ritual occasions.

In African societies, older people are respected, and status within society comes automatically with age. The head of a group is usually the oldest male member. They tend to be gerontocracies, and advancement up the chieftaincy ladder may depend on the decease of an incumbent, whereupon all those below him move up one step. Not surprisingly such societies are conservative rather than radical; but they are also stable. However there are some peoples, often living in dispersed rather than nucleated communities, such as the Igbo of eastern Nigeria who have no system of inherited chieftaincy, or the Bamileke of Cameroon where rank and titles can only be won by strength of personality and achievement within society, where young men could be treated as elders. Such peoples often lacked the organisational power of more formalised communities, but when suzerain control was imposed by an outside power, in the first instance by colonial government, they responded by becoming highly

4 Seton, R.S. quoted in Wilson-Haffenden, J.R. (1930) *The red men of Nigeria*, Seeley Service; reissued by Cass, 1967 (p. 318)

progressive and successful groups. Sometimes they pushed their luck too far, as may have happened during their respective civil wars.

African traditional farming

The allocation of land for farming is usually in the hands of older men. The chief of the community, or the family head, acts as trustee in the disposal of land. Land can be allocated only with his approval, and he cannot allocate it without at least the tacit consent of the majority of the community. Because land is worked cooperatively with women, a wife is an essential adjunct to farming. A man must pay the bride price before he can acquire a wife. This commonly exceeds a year's wages of a government labourer, though among the Yoruba and Ashanti, for example, it is nominal. Although this onerous expense is a source of bitter complaint, it is a part of the marriage contract and a guarantee of the stability of marriage. If a man deserts his wife, he loses his investment. If she deserts him, the children continue to belong to him and will remain in his custody, and she may not remarry unless the dowry is repaid. In the event of estrangement, the wife's parents may put considerable pressure on her to remain with her husband, because of their liability to refund the dowry.

In the forest zone, mixed cropping is usual, and the various crops such as yams, maize, melon, amaranthus, sugar cane, groundnuts and peppers are intermingled, though each is planted and harvested at its appropriate time. In unfertilised fields this gives greater production per unit area than does monoculture, because of the more thorough exploitation of the soil by the roots of the different crops. This is important, because the maximum productivity must be obtained from the limited area that a man can clear of trees and woody growth. The man does the heavy work, such as felling trees and ridging or mounding the land, and the root crop (such as yam) belongs to him. His wife does the hoeing and weeding, and the vegetables and peppers belong to her. Any surplus is sold and, though women are normally responsible for marketing food crops, the revenue belongs to the respective owner. There may be regional variations; for example cassava may be a women's crop, because women make it into *gari* or *fufu*, but in areas where the land has become too infertile for yams, as in parts of eastern Nigeria, it may become a men's crop, though the men may share it with their wives (who convert it into *gari*) or sell the crop in the ground to women who make and sell *gari*. Crops are normally ascribed either to men or to women.

Working on the farm with a matchete

Women and children work on the farm too

In a polygamous society a woman is expected to produce enough food to feed at least herself and her own children. In Africa it is usually assumed that a woman should earn her own keep, and consequently women are more economically independent than in Europe. In Cameroon and Central Africa, women do nearly all the farmwork, though they would not be expected to clear mature forest for themselves. However, in the forest areas of West Africa, farmwork is shared more equally between the sexes. In the savanna zones, men do most of the farming. This is sometimes explained as being due to the prevalence of the Moslem religion among savanna Africans, and the practice of purdah, though in fact seclusion of women was formerly little practised by peasants in sub-saharan Africa. A more probable reason may be the lack of security for women working alone in open savanna country during the disturbed times immediately prior to European colonisation. Moreover, in savanna regions, the staple foodstuffs are cereals, and women have to spend considerable time pounding them with pestle and mortar to grind them to flour. Also women spin, weave and dye cotton to make clothes – though in some regions (e.g. Yoruba) both men and women may weave. Among the Hausa it is the men who practise embroidery.

Farms are normally worked individually, even though the land may be owned communally. However, a group of men may dig one another's fields in turn, or a group of women may hoe each other's fields. This may be an act of charity to help a neighbour in difficulties, due to sickness or absence, or it may be a customary practice. It is known as *aro* amongst the Yoruba, and *gayya* amongst the Hausa. The participants do the same task at the same time, competing against each other to boost their productivity. The work may be accompanied by drumming, and the beneficiary will be expected to provide food and perhaps beer or wine, so that drudgery is turned into a semi-festive occasion. This brings to mind similar communal enjoyment from hard work during the harvesting of wheat and other cereals in Europe, before mechanisation.

Traditional rights to tree crops

It has been an almost universal custom to retain useful trees in the farm because they yield foodstuffs, fibres or medicines. For example in the forest zone, trees such as *Chrysophyllum albidum*, *Dialium guineense* and *Irvingia gabonensis*, which provide sweet edible fruits, will be kept, as will other trees such as *Pterocarpus osun* (for the leaves), *Pentaclethra macrophylla* (oil bean) and *Treculia africana*

This woman is pounding cassava

(African breadfruit). This may well concentrate the proportion of useful trees in secondary forest. In savanna areas, examples of trees preserved in farmland are *Butyrospermum paradoxum* (Shea butter), *Adansonia digitata* (Baobab, edible seeds, bark for fibre, young leaves as vegetable), *Tamarindus indica* (edible fruits), *Parkia biglobosa* (African Locust Bean), *Acacia nilotica* (fruits for tanning), and *Acacia albida* (fruits for fodder). The extent of this practice varies. The adjacent lands of the Tivs, a savanna people, and the Ogoja, a forest people, may be distinguished by their respective effects on the landscape. The Tivs remove nearly all the trees from their farms, while the Ogoja retain them. However cash-cropping, rather than traditional subsistence farming, has tended to eliminate trees. This is true in both savanna and forest zones. Excessive removal of trees is harmful, because they help to stabilise the soil and to restore soil fertility; in arid regions they restrain the force of the wind, preventing sand drift and alleviating the reduction in crop growth that results from high evapo-transpiration.

The *rights of usufruct*, to farm trees of naturally occurring species, do not necessarily belong to the farmer who cultivates the land on which they stand. In some communities these belong to the proprietor

Traditional Hausa farmland

of the land, and in others to the whole community. Lloyd, referring to Ondo Province of western Nigeria, writes that palm fruits belonged to the farmer while he was actually cultivating the land around the trees, while at other times they could be collected by any member of the community.[5] Chubb describes the practice in Aba Division of eastern Nigeria, where on a certain day in every twenty days, a drum is beaten and every member of the community may cut as much produce from the communal palms as he desires and can manage to do during the day allotted.[6]

In the case of palm trees, the men harvest the fruits, but the women extract the oil. Traditionally the palm oil belonged to men, who sold it for export though some was retained for cooking, whereas the palm kernel oil belonged to women who used it to beautify the skin or as medicine and, with the potash obtained from burning the palm bunch refuse, made soap. The crushed fibre from the fruits is dried and used as fire lighters.

5 Lloyd, P.C. (1962) *op. cit.*
6 Chubb, L.T. (1961) *Ibo land tenure*, Ibadan University Press

In Ijebu-Ode Division of western Nigeria the fees paid to exploit timber trees outside forest reserves are divided between the *oba* (king), the village chief and the farmer on whose land they grow. These last two may receive their payment either as cash or in kind (as planks from pit sawn timber). In other States of Nigeria the revenue may be paid wholly to the local authority. In Nigeria, the Forestry Ordinance of 1908 introduced the concept of 'protected trees'. These were trees of listed species which yielded useful products, whether fruits or timber. Forestry Department permission had to be obtained before felling or destroying these trees. The main consequence was to add to the unpopularity of the Forestry Department, because of the scope for abuse of power that it gave to Government Forest Guards; and it also discouraged farmers from making any effort to foster the natural regeneration of these species, which effectively no longer belonged to them. Certainly the numbers of farm trees in the traditional farmlands that surround the cities of northern Nigeria have continued to diminish. The law appears to have been counterproductive, and to have been used mainly to generate local authority revenues. It would have been much better to have tried to achieve the same purpose by public education and enlightenment.

Trees that are often also left standing are the ironwoods or 'break-axes', though in some cases (such as that of *Cylicodiscus gabunensis*) they may burn standing, and can be destroyed by lighting a fire round the base of the trunk. Farmlands are the most important source of firewood, by collection of the dry wood remaining from clearing the farm. In fact natural forest contains little useful firewood. It is the women who normally cut and carry home firewood for domestic use, and they may have to carry it several kilometres. Where men cut firewood, it is usually for sale. Women are also responsible for carrying water to the house, which may be from a considerable distance. Men are reluctant to do tasks regarded as women's work, and vice versa. Moreover some work is considered demeaning for older people, for example bird-scaring by men, and sweeping the house and carrying water by women. These will be performed by younger members of the household.

Compound trees normally belong to the persons in whose compounds they grow, that is the land immediately surrounding the dwelling house. Indeed the ownership of a tree is usually recognised as belonging to the person who planted it, though obviously there are difficulties in maintaining ownership over trees of naturally occurring species, which cannot easily be distinguished as having private ownership. In forest areas the planting of orchards has probably resulted from the introduction of exotic tree species as cash crops,

A cocoa plantation

such as citrus, cocoa, coffee and para rubber (in the last quarter of the nineteenth century) – and also Cola, in particular *Cola nitida* introduced from Ghana to Nigeria in about 1900. The deliberate planting of perennial crops has effectively changed land from communal to individual ownership, and this has been a general consequence throughout the forest zone of West Africa. Tree crops usually belong to men, though among the Yoruba, women may also acquire tree crops. If women's labour is needed to harvest a men's crop, the women will expect to be recompensed in cash or kind or with a part of the crop.

Traditional responsibilities of men and women

Besides mutual benefits, men and women have mutual obligations. Because of the nature of men's crops, men's earnings tend to be in lump sums at the time of harvest, so the men are responsible for such large disbursements as school fees or rebuilding the house. Women's earnings come in frequent small amounts, so women are expected to meet the running expenses of the household. Girls have tended to be neglected educationally, though the effect on the bride price may influence the decision of whether and how far to educate them. This discrimination may occur because men are expected to contribute more to the general benefit of the community than are women, who tend to concentrate on the welfare of their own children. Women continue to belong to the family into which they were born, and not to that of their husbands, and traditionally may be returned to their own families after death. In fact the bride price is paid in order that the right to possess the children born to her passes to the husband's family. Otherwise the children would continue to belong to the wife's family. It may not be a coincidence that among peoples where the bride price is low, inheritance of land takes place through the maternal line (not the male), as is still the case for the Ashanti, and was formerly amongst the Yoruba. Fadipe quotes Johnson as recording the change from inheritance of a man's property by his siblings to that of inheritance by his offspring, as being decided at Ibadan in 1858.[7]

Polygamy is advantageous for this system of agriculture, provided the husband can clear sufficient land for all his wives. The enlarged area of cultivated land under the control of one man increases the security of the whole family. However women commonly resent their co-wives, and may limit their food production so as to grow sufficient only to feed themselves and their own children. Therefore a family does not produce in proportion to the number of wives.

Okediji (in Fadipe) states that, 'although polygyny was valued in traditional Yoruba society for social and economic reasons, it did not appear to be the preferred form of marriage for many except those who were relatively affluent.'[7] The writer's memory is that during the 1950s in the rural Muslim north of Nigeria, among Fulani and Hausa peoples, the possession of more than one wife was not usual. A reason is the problem of paying the bride price: the efforts of father and son may have to be joined in order to gather sufficient wealth to meet this expense, so that the son may be married. A son would not be content to see himself deprived of a wife, and the status which marriage confers, in order that his father could acquire another wife! However, marriage is essential for the functioning of peasant society, because of the division of labour between the sexes. A single man or woman can rarely farm either efficiently or extensively. Where a wife becomes incapacitated or dies, a man will have to obtain another. Sometimes a substitute is provided by his wife's family, without a further dowry payment. He may, of course, pay by instalments, but the wife will not

7 Fadipe, N.A. (1970) *op. cit.*

transfer to his compound until a substantial part has been paid.

Men and women are born in approximately equal numbers, but in rapidly increasing populations there tend to be more nubile women than men, because women tend to marry younger than men. For tribes with a high bride price, the discrepancy in age may be considerable because of the problems of amassing the necessary funds. This is frequently alleged to be a cause of immorality and prostitution, because men cannot afford to pay the bride price, and young women are denied the opportunity of marriage.

Deficient medical services in rural areas cause appreciable losses of women of childbearing age, due to deaths during childbirth or related gynaecological problems. A widow was inherited by a younger brother or a son of her former husband, though a marital relationship would not extend to her own son, or if she were beyond childbearing age. In the villages women tend to be in surplus because of the greater migration of the men to the towns, and perhaps in pre-colonial times because men were lost by internecine warfare or slavery.

A peasant farmer and his family themselves consume between a quarter and three-quarters of their total production. Peasant farmers appear to have considerable free time during much of the year, though they may be overworked during periods of land preparation or of harvest. This was necessary, because traditional communities were relatively isolated, and had to satisfy most of their own needs by making clothing, utensils, furnishings, tools and shelter for themselves, besides producing their own food. Each small community had its own blacksmith, wood-carver, calabash maker, leather tanner, mat-maker, spinner, weaver and dyer of cotton cloth, and potter. (The potter was usually a woman.) There would be men who specialised in the construction of house walls and roofs, though most people would possess more than one skill in which they could take pride. A considerable amount of time was also spent on social and religious observances. Though apparently relatively pleasant and unstressful, it does not appeal to economists as a productive use of time. Nevertheless there is a considerable element of uncertainty and risk for the household head, which may explain why people prefer regular salaried or wage-earning employment, even though the work may be more arduous and tedious, and not particularly well-paid.

Obstacles to improved farming

Peasants are not wholly reactionary and unprogressive. Like gardeners round the world, they observe their neighbours and seek from them new crops and new varieties. Once an idea is accepted it can spread rapidly through the community. This happened for example for cocoa growing in western Nigeria, despite the lack of government farming extension services at that time (see *Table 3*). Nevertheless drastic changes to the farming system can be difficult to make.

Though the key to permanent agriculture in savanna areas may well be mixed farming, the Anchau scheme begun in northern Nigeria in the 1940s, which attempted to introduce the principle to shifting cultivators, had failed fifteen years later, and weeds were growing over abandoned ploughs and harrows and farm buildings. A contributory factor was that traditionally in this area the cattle-keeping people were nomadic Fulani, and the Hausas, who were the sedentary arable cultivators, were unable to keep alive the beasts provided for them. Nevertheless the settled Fulani of the Gombe area, who both kept cattle and practised arable cropping, have been using ox ploughing successfully since the 1950s. Once the idea was

Table 3 The development of cocoa farming in Ghana and Nigeria: cocoa exports (thousands of tonnes)

Year	**Ghana** (Gold Coast)	**Nigeria**
1886	·	0.0012
1890	·	0.0061
1895	·	0.0215
1900	0.5	0.0920
1905	5.1	0.47
1910	22.6	2.93
1915	77.3	9.10
1920	125	17
1925	218	45
1930	191	52
1935	269	88
1965	572	294

Data obtained from:
Hill, P. (1956) *The Gold Coast farmer,* OUP (p. 139)
Berry, S.S. (1975) *Cocoa, custom and socio-economic change in rural western Nigeria,* OUP (p. 240)

Ox ridging

introduced to them, and the ploughs made available for purchase, the method sold itself.

More recently, near Yaoundé in Cameroon, in the forest zone, peasant farmers demanded that an extension project (then in its seventh year) should be abandoned, even though it had been demonstrated that the use of improved methods could increase yields by perhaps 60 per cent. On the initiation of the scheme, fertilisers and other inputs were provided free by the technical aid agency. When they were shown to be successful it was hoped that the peasants would buy them for themselves; but the people preferred to return to traditional ways and not be pestered by extension workers. A reason is that subsistence crops cannot pay sufficiently for inputs that have to be bought, because a substantial proportion of the crop is not sold, unlike cash cropping. Moreover the purchase of inputs adds a pecuniary risk to the natural hazards which a peasant farmer is already incurring.

Another disincentive is that peasants found that, although the use of fertilisers gave an appreciably greater yield of yams, their keeping properties were impaired, a fact which had escaped the agronomist. The attitude of the peasants to a drastic reduction in the supply of a staple foodstuff, as a result of rotting in storage, may be imagined.

The problem is said to be due to the lack of a balanced fertiliser addition.

One may ask whether it is desirable that successful farming should become dependent on managerial skills, on expert guidance and on inputs that cannot be sustained after the end of the technical aid project – when the fertilisers or agricultural chemicals can no longer be obtained, when the tractors or the pumps break down and neither the spare parts nor the mechanics are available, when the country cannot afford to import sufficient animal feed, when the government extension workers have run out of travelling money ... Perhaps peasants and small farmers should not be 'bribed' into what are experimental systems of farming. They are in no position to recover from bad mistakes, whatever the cause. The survival of a family, and indirectly of the community on whom they may otherwise become a charge, depends on the certainty that at least enough food will be produced to last them through the year. A temporary failure cannot be offset by borrowing from the bank, as perhaps it could be by farmers in a developed country. Fortunately peasants are not easy to dragoon into ways that, to an outsider, appear to be for their benefit or for the benefit of the nation. If they feel they are exploited they can more or less opt out of commercial activity and still live from their own resources – and are quite prepared to do so, as the decline in cocoa production in West Africa clearly demonstrates.

Nevertheless peasants are not insensitive to economic factors. Both groundnut production round Kano and cocoa production round Yaoundé were developed in opposition to government policy at the time. Around Kano the then colonial government had intended to promote the growing of cotton; around Yaoundé the intention had been to discourage smallholders from growing cocoa, which was to be produced instead by large expatriate owned plantations elsewhere in Cameroon. During the 1970s, cocoa production in Cameroon, Ghana and Nigeria slumped, due largely to an unsatisfactory price to the farmer and to difficulties in obtaining payment, not unassociated with the activities of government marketing boards. In Cameroon, production of robusta coffee was rising, while that of arabica coffee fell. It is of interest that arabica coffee was subject to the controls of a marketing board, whereas robusta was not, but was directly purchased by commercial firms.

In Midwestern Nigeria, smallholdings of rubber are being clear-felled to make way for arable crops, because of the low price of rubber and the difficulty of obtaining labour for tapping. At Zaria, a factory for manufacturing tins of tomato purée was constructed, and peasant farmers were provided with a package to grow tomatoes. This proved very successful, and numerous farmers turned to growing tomatoes. Unfortunately it proved more profitable for them to sell their crop as dessert fruit, and eventually the factory had to close down because of a lack of supply of tomatoes! Because of the small area farmed, scientifically improved varieties and methods of cultivation still generate relatively small incomes. Small wonder that farmers prefer to invest their savings in the education of their children, rather than in the development of land.

Modern influences on farming and society

It is of interest to quote Mellanby's observations on the impact of education on agriculture; 'One criticism of "education" that is heard all over Africa is that up to now it has been responsible for removing people from the land. It is alleged that as soon as the son of a peasant farmer gets a smattering of book learning, he feels he is too superior to dig in his farm, and he therefore tries for a "white collar job". His training is perhaps insufficient to fit him for such, or no opening may exist, and then there is the danger that he will become just another unemployed, and perhaps unemployable, member of a landless proletariat. All this is, in its way, true. Education *will* make the son of a small farmer unwilling to live his whole life scratching a bare subsistence. But the question is whether or not this is a good thing. So long as the greater part of Africa is divided up into pocket-handkerchief sized holdings, which incidentally do not give anything like an outlet for the whole energy of even a moderately strenuous individual for the whole year, so long will progress in agriculture as well as in other walks of life be retarded. Only by causing a substantial proportion of farmers to leave the land to be worked by a more efficient minority will farming be improved. It can well be argued that this will not make for more happiness. The peasant in the tropics . . . had a very easy life, and if his standard of life was low for much of the year he had a sufficiency of food and what to him was adequate shelter and clothing . . . But even if we take this somewhat idyllic view of the life of a peasant farmer in the tropics, for better or for worse (and I believe that it is definitely for the better) the situation is changing with increasing rapidity. Food supplies for the rapidly increasing population are endangered as more and more workers are leaving the land, and it is our job to see that this leads to improved conditions of life for the country and not disaster.'[8] Oyenuga writes,

8 Mellanby, K. (1958) *op. cit*.

'So long as agriculture in Nigeria and other parts of West Africa continues to offer no other equipment than a cutlass, a hoe and an axe for no other reward than merely to keep body and soul together, so long will it continue to offer no attraction to the primary or medium school-leaver, or even to an illiterate young African as a means of livelihood; and so long will the army of the unemployed continue to mount in the towns and cities.'[9]

Young people want to escape irksome obligations to their elders. Young men want wage-paying employment in order to have money in their pockets and ultimately to be able to pay the bride price. In commerce the possibilities are infinite, whereas the earnings from peasant farming are decidedly circumscribed. Thus there are strong inducements to leave the village to obtain paid employment elsewhere. If a man does not find a well-paid job, he may return to farming in middle age when the cost of housing, feeding and educating a growing family weighs most heavily upon him. Another incentive to return to farming is that, on the death of his father, he risks alienation of himself and his children from the land if he does not return to claim his patrimony. In other cases, the wife and children may return to the village to farm, while the husband continues to work in town and hires labour to fulfil his obligations to the household. The cash he earns helps to meet the expenses of his family and near relatives for medical treatment and education, and to meet his other obligations to the community. Moreover to return is an admission of failure.

Busia writes, 'With the new opportunities for making a living, the bonds of kinship are no longer as advantageous as before, and are, indeed, a burden to the more enterprising members of a lineage. Many traders and literate young men prefer to work away from home in order to escape the importunate demands of their kinsfolk for financial assistance. A woman trader who has left her home to go and live at Obuasi said she had to leave home because her kinsfolk bought her things on credit, and never paid her. She could not very well sue them in court for the money because they were her relations. In consequence, her business was failing, and she finally decided to leave home. Many literate young men have similarly left home, or when in the Government employment have begged not to be stationed in or near their homes because they are unable to save money there, and in fact often run into debt when they are near home owing to the demands of their kinsfolk for money for subsistence or clothing or the expenses of a funeral or litigation. So the ties of kinship are weakening

9 Oyenuga, V.A. (1967) *Agriculture in Nigeria*, FAO, Rome (p. 308)

because certain types of economic pursuits do not succeed in the framework of the close kinship ties.'[10]

Decay of traditional society and farming systems

Agricultural extension is founded on the assumption that peasant farming can be improved progressively until it is transformed into more productive and profitable forms of agriculture. Unfortunately the experience in Europe suggests that, for a traditional society, the consequences of the interdependence of the social system with the farming system is that changing one invariably disrupts the other. In Africa both are breaking down and there is mutual degeneration.

In shifting cultivation, all that the peasant may invest in the land is his own labour. He does not usually improve the soil by manuring, by subsoil drainage or by irrigation, as might be found in Asia or in Europe. Exceptions are the terraced farms of former pagan villagers on their hilltop eyries in northern Nigeria; but since the cessation of the 'Time of war' these are now often abandoned in favour of shifting cultivation on the plains. Investment of capital in agriculture depends on its being sufficiently profitable, which in turn depends on the development of systems of permanent agriculture, and of land tenure reform to disencumber it of communal rights.

Finally there is the need to introduce into farming industrial concepts of management and production, together with commercial methods of accounting and financial control. In Europe, businessmen have been encouraged to enter farming and forestry, initially by the considerable improvement in land values resulting from introducing improved methods of farming, and more recently by grants, subsidies and tax concessions. These have been 'new men with new ideas'. A consequence of more profitable farming is a rise in land values and rents, and thereby a necessity for sufficient investment by farmers. High land values encourage the poor farmer to sell his land, and force purchasers of land to farm it efficiently. High rents have the same effect, in that only efficient farmers can afford to pay them. Farmers who are inadequate, for whatever reasons, are pushed off the land. They must become businessmen or they will not survive.

The agricultural problems of Africa can only be solved by innovative farming, and this requires innovative farmers. It needs an

10 Busia, K.A. (1951) *The position of the chief in the modern political system of Ashanti*, Oxford University Press (p. 233)

Traditional terraced farms in Northern Nigeria

audience of farmers who are receptive to new ideas and to experimental findings, and who are capable of putting them into effect. Peasant-scale farming cannot justify the expense of educating highly trained farmers, nor can it reward them adequately. This is why the response of the agriculturist in tropical Africa to the appeal to grow more food is to cry out for extra money for government research and extension work. At the moment this is his only prospect of sufficient employment and earnings. In England, the key to permanent agriculture depended on the technical breakthrough of learning to use the fallow, which was pasture. This depended on acquiring the ability to keep large numbers of cattle through the winter, by the use of fodder crops, and on discontinuing communal land ownership, by dividing the land and fencing the fields. In Africa the rationale will be the same, though the remedies may be different; these are discussed in later chapters of this book.

Peasant society contains much that is satisfying, but is already undermined by increased populations and by a desire for better living conditions. Alternatives are needed that will satisfy both economic and social needs. For most people this is probably not farming.

There will be a niche for the smallholders as market gardeners round the towns, using urban waste as manure, but this will be to produce high-value specialist crops rather than staple foodstuffs. Urban growth is unstoppable, but should be encouraged to spread to other parts of the country rather than to concentrate itself round a few cities that become overwhelmed by uncontrolled expansion. It is very desirable to do whatever is possible to improve the living standards of ordinary people, but social development should not be confused with agricultural production. Though the earnings of peasant farmers ought to be enhanced, improvements in their earnings may, paradoxically, actually hasten their departure from the land by providing the means to do so. Boosts to peasant agriculture can be regarded only as a palliative to tide the nation over while business farming develops.

In the villages of southern Nigeria, despite free primary education and new schools, despite heavily subsidised fertilisers, despite rural electrification and water supply, the houses have been falling down and not being rebuilt – even in areas in which the soils are naturally fertile and which are near large towns. There is every reason to suppose that in perhaps a decade, the contagion will have spread throughout the north of the country, which has always lagged behind the south in its social development. We may have to learn to live with the fact that peasants are a disappearing breed.

3 The biological impediment

> *'Within five years of constructing and laying out the Institute's research farm according to the most scientifically sound Western technology, the land was deteriorating fast and crop yields were declining . . . The fancy terraces and contour bunds didn't work out as expected on the fragile tropical soils. [We were] faced with a visibly embarrassing example of how not to farm in the tropics. Now, of course, we know that when the bush is cleared for cultivation, the soils change for the worse rapidly and irreversibly. The soil becomes less permeable and more vulnerable to erosion.'*
>
> *Ermond Hartmans*

Agriculture is applied ecology. If a farming enterprise strays beyond the ecological limits it will lose money; at worst it may go bankrupt, and the invested capital would be lost. This could be a consequence, for example, of irreversible damage done to the soil.

Soil is the product from the interaction of climate, vegetation and geological parent material over long periods of time. It may take decades, possibly centuries, for a soil to develop naturally; yet the fragile soils of the lowland tropics can be swept away in as many months, to leave a barren and unproductive landscape. Without soil, higher plants can barely grow; and to restore devastated terrain is an extremely expensive process, as those responsible for mining reclamation will realise. The soil is the main resource, and if an enterprise is to thrive it must be a thinking organisation, responding to the natural and economic forces that act upon it. Provided that harmful developments do not occur too quickly, and provided that the organisation can adjust to these forces and does not wander too far outside the ecological limitations, it can remain inside their confines and still survive.

The soil

The soil is the matrix in which plant roots grow and obtain anchorage, and from which they obtain a source of water and supply of nutrients. Roots can absorb nutrients against the diffusion gradient, and water against the osmotic gradient, but for this they need energy and therefore to respire. Thus soil must have a structure that allows it to be permeated by air. Plant species that habitually grow in waterlogged soils make special provision for root respiration, for example by the production of breather roots (pneumatophores) that project above the surface of the mud, as we may see in mangroves and in some freshwater swamp species. Compared with soils of temperate regions, tropical soils tend to be shallow with the greater part of the fertility in the upper 20 cm of topsoil. Under high forest, this contains a dense weft of roots – the feeder roots. In the subsoil, roots are only sparsely distributed, though trees may send sinker roots down to reach the water table. These sinker roots help to make the tree stable, and allow it to have access to water throughout the year, even during the dry season. Trees with pronounced buttresses, such as *Triplochiton scleroxylon*, may also possess strong sinker roots or tap roots.

The ability of soil to hold both nutrients and water depends mainly on two constituents: humus, which is soil organic matter, and clay, which is colloidal mineral particles of aluminium silicates. In tropical soils, generally, the preponderance of clay is kaolinite, which has poor nutrient-retaining and water-yielding characteristics compared with soils of temperate countries, where the clays are mainly montmorillonite. Kaolinitic clay has only ten per cent of the nutrient holding capacity of montmorillonitic clay. A consequence is that, in tropical soils, the humus content accounts for ninety per cent of the soil's ability to hold fertility under high forest, and about eighty per cent under savanna woodland. This is called the **Cation Exchange Capacity** (or CEC) of the soil.

Humus is a brownish or blackish substance, which accounts for the colour of topsoil. It forms from the products of decay from dead plant or animal matter that fall onto the soil, or are incorporated within it. Humus is a complex colloid, preserved by tannins, that breaks down relatively slowly. It has a sponge-like capacity to retain water, and gums soil particles together to give a crumb-like structure. This helps to make soil permeable to both air and water.

Mineral nutrients are also adsorbed onto the colloid; the total nutrients held by the soil as a proportion of the CEC, is known as the **base saturation**. These mineral nutrients, or bases, include calcium (Ca), magnesium (Mg) and potassium (K). Humus and clay may be bound together as stable complexes, due especially to polyvalent cations, such as aluminium (Al), that are present particularly in tropical soils. It is apparent that the humus, or organic matter content, is a very important component of soil fertility in the tropics.

The predominance of kaolinitic clay is a result of soil-forming processes in the tropics, because of higher temperatures and greater leaching of tropical soils by rainfall. Particles of kaolinite have a 1:1 crystalline lattice structure, the constituent sheets each consisting of one layer of silica and another of alumina; they are plate-like and non-expanding, and the negative charges are borne only round the edges of the plate. Particles of montmorillonite are much smaller and have a 2:1 lattice structure, the constituent sheets each consisting of two layers of silica sandwiching one layer of alumina; they are expanding, and negative charges also occupy positions between the sheets as well as round the edges. It is the negative charges that attract the positively charged cations of the soil bases, e.g. calcium, magnesium, potassium, sodium etc. Thus montmorillonite clays are able to adsorb much greater numbers of cations than kaolinitic clays. Water can also be absorbed between the plates in montmorillonite, and in consequence they swell and contract as they are wetted and dried. They are therefore to some extent self-tilling.

In temperate countries, the frost in winter freezes the water in the clay and splits the clods of clay on the soil surface to fragments. These physical characteristics of clays explain why soil in tropical regions is an inferior growing medium for plants, compared with soils in temperate countries.

Solutions that respond less to additions of acid or alkali than pure water would are said to be **buffered.** This is a result of the presence of the salts of weak acids in the solution, such as carbonates, phosphates, silicates and humates – the latter being due to the humic acids from soil organic matter. These salts mop up hydrogen ions in the solution to give undissociated weak acids, which reduces the hydrogen ion concentration and hence the acidity of the solution. This is of great importance in the soil. Soil colloids, that is clays and humic materials, absorb metallic ions and behave like weak acids, exerting a buffering effect on the soil. Sandy soils, which contain only small quantities of buffering substances, tend to be much affected by increases in the hydrogen ion concentration.[1]

1 (a) The degree of soil acidity, or alkalinity, is represented by pH, the decimal logarithm of the reciprocal of hydrogen ion concentration. Most soils are within the range pH 4 to pH 8, from the most acid to the most alkaline as the pH value increases. A value of pH 7.0 represents neutrality.
(b) The cation exchange capacity (CEC) of montmorillonitic soils, typical of temperate regions, is substantially independent of soil pH; but for kaolinitic soils CEC is affected by pH and, for the yellow highly leached soils of the humid tropics, the correlation between CEC and pH may be very strong. Raising the pH, that is making the soil less acid, raises the CEC and lowers the base saturation. The addition of salts (such as artificial fertilisers) to the soil brings iron and aluminium into solution, and results in a lowering of soil pH, that is an increase of soil acidity.

Besides providing anchorage for roots, and acting as a storehouse of fertility, soil is the factory in which organic matter from dead animals and plants is decomposed, and in which chemical reactions break down mineral particles from the parent rocks, mainly by the activity of soil bacteria. This restores mineral nutrients to the soil and makes them available for absorption by roots, and hence to sustain continuous plant growth. About half the available soil bases held in the **biosphere**, that is the air, vegetation and soil together, are found in the leaves, twigs, branches, boles and roots of the trees in the forest; most of the remaining **available nutrients** are in the 20 cm depth of topsoil.

When plant remains fall on to the soil surface and decay, the nutrients are returned to the soil. Indeed it has been said that tropical forest lives on the products of its own decay, because the soil has been deeply weathered and not much in the way of nutrients is available from the parent rock. On the other hand, nitrogen is the principal constituent of air (75.5% by weight) and, though it may be released by the decomposition of organic matter as nitrates for reabsorption by roots, it is also fixed from atmospheric nitrogen by certain bacteria free-living in the soil (e.g. *Azotobacter, Beijerenkia* and *Clostridium*). There are also other nitrogen fixing bacteria (*Rhizobium*) which live in nodules formed on the roots of certain species of plants and trees, in particular those belonging to the plant group Leguminales, but excluding the family Caesalpiniaceae. Some other tree species, for example *Alnus* and *Casuarina* species, which belong to other families, also form such root nodules.

Furthermore, nitrogen-fixing bacteria may be present in the root cortex of some species, such as the grasses *Brachiaria humidicola* and *Panicum maximum* (Guinea grass), and may fix up to 25 per cent of the plants' needs.[2] Fixation of atmospheric nitrogen can proceed rapidly in tropical soils, making a very significant contribution to its availability as a soil nutrient.

Although nitrogen is lost to the soil when woody vegetation is burned, nitrogen fixation in the soil may be promoted by the heat of the fire causing partial soil sterilisation, due to the burning of dried leaf litter, or the cover of dead grasses, or by burning of woody vegetation when the land is clear felled and burned preparatory to cultivation. Another cause of sterilisation may be drying out of the soil, for example during the dry season. A particular effect is a

2 However, grasses are often associated with reduced soil nitrogen levels, and some species have been shown to inhibit soil nitrification.

reduction in the populations of soil fungi and of soil protozoa, which promotes enhanced activity of soil bacteria, and hence also nitrogen fixation. The proportions of soil micro-organisms are also influenced by soil acidity: bacteria thrive best in pH between 6.2 and 7.5. As the pH is decreased, and the soil becomes more acid, fungi increase at the expense of bacteria.

Clear felling and burning the forest causes a rise in soil pH, due to the bases added to the soil from the ash. In other words, burning the vegetation may have a beneficial effect on the soil. Of course other bacteria break down soil nitrates to gaseous nitrogen, or the nitrates may be leached away in the drainage water, and hence lost to the system. In this way, nitrogen levels in the soil may fluctuate. For example they tend to rise shortly after the onset of the rainy season, and to diminish subsequently.

Unlike the soil bases, the amount of nitrogen available for plant growth is not determined by what is already present in the system. This may be contrasted with sulphur, which is lost to the atmosphere when the woody vegetation is burned, and cannot be restored except from the lower soil horizons or by additions of organic manure. In consequence savanna soils tend to be deficient in sulphur.

Phosphorus is relatively immobile in the soil, and roots have to explore to find it. It is fixed in the soil as calcium phosphate, particularly in alkaline soils; and in more acidic soils as hydroxy-phosphates of iron and aluminium. These substances are insoluble. A substantial proportion of soil phosphorus is bound in the soil organic matter, and a certain amount adsorbed on to soil colloids. The availability of phosphorus is greatest between soil pH 6 and 7, and indeed this range of acidity seems the best for ready availability of plant nutrients.

At acidities below pH 5, appreciable amounts of aluminium, iron and manganese become soluble, and may be very toxic to plants. Streams and rivers fed from extremely acid soils, below pH 4.5, may be stained brown with dissolved iron and dispersed humic colloids. Above pH 7, in alkaline soils, iron may become fixed as insoluble ferrous hydroxide, and unavailable to plants. Potassium is one of the soil bases, and is held in an available form by the soil colloids – both clay and humus. It is released slowly by the decomposition of soil organic matter, and also by the breakdown of soil minerals. Savanna soils usually have good reserves of potassium, but it may be deficient in forest soils particularly under high rainfall.

Nitrogen, phosphorus and potassium (N, P and K) are the principal plant nutrients that are added to the soil in artificial fertiliser. Lime is a source of calcium and magnesium; because it is alkaline, it may improve the pH of very acid soils and hence decrease aluminium toxicity and improve the availability of the various plant nutrients.

Effects of soil organisms

Despite the boost to soil fertility that seems to be caused by semi-sterilisation and destruction of soil fungi, some fungi are useful in the soil. They are able to decay wood, and in particular its constituent **lignin**. Their ramifying hyphae can penetrate wood and vegetable matter in order to rot them. Some fungi form **mycorrhizal associations** with tree roots, and are present in the living root cortex of almost every tree in tropical forests, though their role is not well understood.

For pines, ectotrophic mycorrhizae coat the outside of the roots, and seem to take the place of root hairs. Their presence is essential for the survival and growth of pines, which seem able to thrive on soils that are too infertile or too skeletal to support other tree growth. If the soil receives large amounts of organic matter that is high in carbon content but low in nitrogen, there will be an increase in the population of soil micro-organisms, which draw on the reserves of nitrogen in the soil in order to multiply. This leads to a temporary drop in available soil nitrogen, which will check plant growth. This is the reason why compost should be decomposed thoroughly before being added to the soil, so that the carbon will have been converted to relatively stable humic materials and also lost to the air as carbon dioxide (CO_2). It is another reason why a farmer burns the vegetation when making his farm, instead of digging it into the ground.

Besides the soil micro-organisms, there are also the soil **macro-fauna** – that is the animal life in the soil that is more or less visible to the naked eye. Insects, mites and small burrowing mammals, such as mice, play an important role, although their effect is largely, but not entirely, mechanical. They create pore spaces in the soil, assisting aeration and drainage, and break up vegetable matter, making it accessible to bacteria and fungi. The most essential are perhaps the earthworms and termites.

Soil passes through the gut of earthworms and termites; earthworms have been shown to bring up about 50 tonnes of soil per hectare per year, and termites about 1½ tonnes. Both creatures carry dead organic matter down into the soil. Termites are able to chew up and digest wood, thanks to the protozoa and bacteria present in their digestive tracts. They help to maintain and restore the fine fraction of the soil into the upper layers, and have been found to have an appreciable improving effect on the acidity (pH) and cation exchange capacity (CEC) of the soil, and may be regarded as natural soil cultivators. They are very active under forest, as evidenced by the carpet of worm casts. However their populations diminish markedly under arable cultivation, though they may be encouraged by soil mulching.

Termites can damage cultivated crops, perhaps especially if deprived of dead vegetable material to consume. Earthworm activity may be enhanced under plantations of certain tree species, such as Teak and Gmelina. These trees appear to absorb bases from the deeper layers of the soil, and redeposit them in the leaf litter on the soil surface, where they become incorporated into the topsoil. The presence of snails, or snail shells, is indicative of diminished soil acidity and the presence of useful levels of soil bases – especially calcium. In fact, successful growth of para rubber has been found to be more related to total soil bases than to available nutrients (unlike arable crops) perhaps due to the ability of trees to act as **nutrient pumps**, by bringing up nutrients from the deeper soil horizons. Some plant species, such as the dense cover of Eupatorium (*Chromolaena odorata*) on farm fallows, appear greatly to enhance earthworm activity. Eupatorium also eliminates harmful nematodes, which can be a very damaging component of the soil mesofauna, especially to certain crops such as bananas, pigeon pea and tomatoes. Nematodes may make it impossible to grow these crops continuously on the same land. Eupatorium also has the useful attribute of smothering and eliminating the rhizomatous Spear grass (*Imperata cylindrica*) which can be a serious nuisance in farm fallows (c.f. the European Twitch or Couch grass).

Soil formation and vegetation zones

As soil develops from the weathered parent materials, horizontal bands of different colours or different intensities of hue, or different textures, become differentiated. These are called **soil horizons**, and the soil appearance in vertical section is called the **soil profile**. In temperate climates, these horizons develop due to:

- the addition of organic matter to the topsoil;
- the leaching of soil constituents down the soil profile, by drainage water;
- the progressive weathering of the subsoil, and decomposition of the mineral constituents.

The upper **A** horizon is stained brown by organic matter, and is the zone of eluviation or washing out of soil material; the intermediate **B** horizon is the zone of illuviation or deposition of leached materials; and the lower **C** horizon is the soil parent material, or material developed *in situ* from weathering of the parent material. Hard pans may form in the **B** horizon, due to the deposition of iron, cementing the soil together.

In the humid tropics there is an additional soil forming process due to **creep**, or lateral movement of soil, mainly downslope. A consequence is that the **A**, **B** and **C** horizons formed from eluviation (and generally identified in temperate soils) are no longer distinct. Hence these soils can only be represented in three dimensions as a soil **catena**. The term means 'little chain' and describes a sequence of soil profiles which appear in regular repetition in regions possessing regular successions of topographical features – for example in undulating landscapes. In the warmth of the tropics, eluviation may include silica as well as soil cations, leaving a soil composed mainly of sesqui-oxides of aluminium and iron.

Soils developed from some parent materials are inherently more fertile than those from other geological origins. For example, soils formed from basic volcanic rocks, such as basalts or lavas, are rich in montmorillonitic clays and are fertile.

Granites are acid igneous rocks, and the soils derived from them are less fertile than those from basic igneous rocks. Sedimentary rocks are formed by the consolidation of debris from earlier rocks, carried away by the rivers and deposited on the floor of the ocean, or the bottom of lakes or lagoons. In the long history of the earth they may be compressed by pressure and recemented by chemical action. Examples are sandstones, limestones and shales. Some may remain unconsolidated as clays, silts and loose deposits of sand. These may be re-exposed at the surface by subsequent movements of the earth's crust.

The soils formed in this way are, so to speak, secondhand and have suffered weathering and leaching for one or more times round. Consequently they tend not to be fertile. Metamorphic rocks are formed from either sedimentary or igneous rocks, which have been altered by pressure and heat inside the earth (such as marble from limestone, or slate from shale) and may be intruded with molten material to form, for example, **gneisses** or **schists**. The soils derived from these rocks are intermediate in their fertility.

To give a simplified account of the features of West Africa: the rocks of the shield area are typically granites, gneisses and schists of the **basement-complex** laid down during the Pre-Cambrian era; that is they are ancient rocks that predate the geological fossil record. The

Guinea savannah

vegetation of this zone is generally Guinea savanna, which may be differentiated into the drier **Northern Guinea Savanna** and the moister **Southern Guinea Savanna**. Soils on this formation tend to be shallow, with poor access to water. Because they are on the same geological formation, the two vegetation types are not sharply separated. The differences are due more to the duration and severity of the dry season, than to the mean annual rainfall which tends to be similar across the range. The Guinea savanna zone has the rather uniform appearance of **parkland**, with dispersed trees and intervening grass; nevertheless the species composition varies with soil depth and rainfall distribution. Though the species compositions are different, the two vegetation types merge into one another. Brief lists of characteristic species for the Guinea savanna zones are given in *Table 4*, together with species lists for the other vegetation zones. An interesting feature of the Guinea savanna zones is the intergrading sub-speciation of *Combretum collinum* and *Terminalia* species. This puts one in mind of the **hybrid swarms** of *Populus* and *Salix* species in north temperate countries. *Khaya senegalensis* is the typical mahogany of this zone. The typical palm is the Fan Palm, *Borassus aethiopum*, found especially on areas of seasonal swamp; the stem is resistant to

Table 4 Natural vegetation zones and characteristic species in Nigeria

Zone	Geological material	Tree species	Grass species
Sudan savanna	Aeolian sands	*Acacia senegal* *Anogeissus leiocarpus* *Balanites aegyptiaca* *Combretum glutinosum* *Commiphora africana* *(Hyphaene thebaica)*	*Cenchrus biflorus* *Ctenium elegans* *Diheteropogon hagerupii* *Loudetia togoensis* *Schoenefeldia gracilis*
	Low lying sites	*Acacia nilotica* *Diospyros mespiliformis* *Mitragyna inermis*	*Digitaria acuminatissima* *Eriochloa nubica* *Hemarthria altissima* *Panicum laetum* *Sorghum aethiopicum*
	Black cotton soils	*Acacia campylacantha* *A. seyal* *Ziziphus spina-christi*	*Pennisetum ramosum* *Urochloa trichopus*
Northern Guinea savanna	Basement-complex: Dry variable depth soils	*Burkea africana* *Butyrospermum paradoxum* *Cussonia barteri* *Isoberlinia doka* *Parkia biglobosa*	*Andropogon schirensis* *Brachiaria jubata* *Ctenium newtonii* *Hyparrhenia smithiana* *Panicum phragmitoides*
	Shallow rocky soils	*Bombax costatum* *Boswellia dalzielii* *Entada africana* *Isoberlinia tomentosa* *Sterculia setigera*	*Andropogon pseudapricus* *Aristida hordeacea* *Hyparrhenia involucrata* *Loudetia arundinacea* *Schizachyrium exile*
	Eroded indurated soils	*Monotes kerstingii* *Uapaca togoensis*	*Andropogon fastigiatus* *Microchloa indica* *Schizachyrium nodulosum* *Tripogon minimus*

Note: Palms and sedges are listed in brackets

Table 4 continued

Zone	Geological material	Tree species	Grass species
	Deeper soils	*Anogeissus leiocarpus* *Entada abyssinica* *Khaya senegalensis* *Terminalia glaucescens* *(Raphia sudanica)*	*Andropogon gayanus* *Cymbopogon giganteus*
Southern Guinea savanna	Basement-complex	*Daniellia oliveri* *Lophira lanceolata* *Pterocarpus erinaceus* *Syzygium guineense* *(Borassus aethiopum)*	*Andropogon tectorum* *Beckeropsis uniseta* *Hyparrhenia subplumosa* *Monocymbium ceresiiforme* *Schizachyrium sanguineum*
Derived savanna		*Afzelia africana* *Bridelia ferruginea* *Parinari curatellifolia* *Pericopsis laxiflora*	*Imperata cylindrica* *Panicum maximum* *Pennisetum purpuerum* *Setaria anceps*
Gallery forest (forest outlier)	Basement-complex	*Albizia zygia* *Antiaris africana* *Chlorophora excelsa* *Erythrophleum suaveolens* *Khaya grandifoliola* *Diospyros monbuttensis*	*Olyra latifolia*
Dry high forest	Basement-complex	*Cola gigantea* *Entandrophragma angolense* *Khaya ivorensis* *Nesogordonia papaverifera* *Pterygota macrocarpa* *Sterculia rhinopetala* *Triplochiton scleroxylon* *Diospyros dendo*	*Leptaspis cochleata* *Oplismenus burmannii* *O. hirtellus* *Panicum brevifolium*

Note: Palms and sedges are listed in brackets

Zone	Geological material	Tree species	Grass species
	rocky hills	*Bombax buonopozense* *Canthium subcordatum* *Hildegardia barteri* *Sterculia tragacantha*	
Wet high forest	Basement-complex	*Brachystegia nigerica* *Entandrophragma utile* *Klainedoxa gabonensis* *Lophira alata* *Pycnanthus angolensis* *Pausinystalia talbotii* *Strombosia pustulata* *Diospyros suaveolens* *(Calamus deerratus)* *(Raphia hookeri)*	*Guaduella* spp. *Isachne buettneri* *Leptaspis cochleata* *Oplismenus* spp. *(Scleria boivinii)*
	Sedimentary	*Berlinia confusa* *Combretodendron macrocarpum* *Brachystegia kennedyi* *Gossweilerodendron balsamiferum* *Guarea cedrata* *Entandrophragma cylindricum* *Parkia bicolor* *Pterocarpus soyauxii* *Diospyros crassiflora* *(Elaeis guineensis)* *(Ancistrophyllum secundiflorum)*	*Centotheca lappacea* *Leptaspis cochleata* *Oplismenus* spp. *Panicum brevifolium* *Streptogyna crinata*
Swamp forest		*Anthostema aubryanum* *Carapa procera* *Mitragyna ciliata* *Uapaca staudtii* *(Raphia vinifera)*	
Note: Palms and sedges are listed in brackets			

termites, and is split to make house rafters.

In some places, the basement-complex reaches towards the coast, and is covered by natural high forest. The more northerly part of this forest is drier than further south, and is typified by tree species belonging particularly to the family Sterculiaceae, of which *Triplochiton scleroxylon* is the commonest emergent. The scarlet flowered *Hildegardia barteri* commonly grows on rocky hills. The typical mahogany is *Khaya ivorensis*.

Under higher rainfalls, exceeding say 2000 mm annually, tree species belonging to the plant group Leguminales are more frequent, of which a conspicuous emergent is *Brachystegia nigerica*. The crowns of this species flush with bright red new leaves at the onset of the rainy season, as do also those of another characteristic species *Lophira alata* which belongs to the family Ochnaceae. Both these species are extremely hardwooded; *Lophira* is an ironwood and the timbers are used as bridge beams and as dock piling due to their resistance to tunnelling by the Teredo worm. *Strombosia pustulata* is a common smaller tree, the bole diameter rarely exceeding 50 cm, and the wood is resistant to both insects and fungi, and is useful for veranda posts and for constructing walls of mud and wattle houses. The typical mahogany is *Lovoa trichilioides* (African Walnut). Characteristic palms are the Oil Palm, *Elaeis guineensis*, and in swampy areas the Raphia Palm, *Raphia hookeri*; both species are a source of palm wine.

In the south, the rocks of the basement-complex shield area are overlain by sedimentary rocks of the Cretaceous, Tertiary and Quaternary periods, derived from deltaic and marine deposits, and include deep loose sands. These are usually covered by high forest, with species belonging typically to the Leguminales group of families. Where the texture of the soil is loose, tree roots can penetrate to the water table and gain access to water throughout the year. Perhaps in consequence these forests are taller and less broken in structure than those on basement-complex derived soils, and the proportion of evergreen species is higher. Beneath primary forest, the undergrowth is not dense, and it is possible to walk through it without a knife. However there may be climber tangles where a tree has fallen, or where the soil may be waterlogged, or where there is a shallow soil depth.

Because of a high sand content, these soils are often of low inherent fertility; they leach rapidly when exposed and have a poor water-retaining capacity. They are edaphically dry, and it may be possible to burn farms much later into the rainy season, than in places with basement-complex derived soils. Therefore they are readily degraded by farming to **derived savanna**, which is composed of grassland

Moist Sudan savannah

Dry Sudan savannah

invaded by stunted trees of savanna species. The grass is tall, and burns fiercely during the dry season; also, because of the impoverished soil, forest species may have difficulty re-establishing themselves. This savanna is intersected by **gallery forest**, mostly in depressions and along water courses, which form outliers to the main forest blocks. The typical mahogany of these outliers is *Khaya grandifoliola*.

To the north, the rocks of the basement-complex are overlain by loess soils derived from the aeolian sands of the Sahara desert, and formed during geological history by periodic southward migrations of the desert. Sometimes, in more northerly regions, the shape of former sand dunes can be discerned in the landscape. The vegetation on these soils is termed **Sudan savanna**, and on the even more arid areas further north, **Sahel savanna**, though the latter scarcely enters into Nigeria. The typical trees are thorny and include *Acacia* species, *Ziziphus* species, and *Balanites aegyptiaca* which is called *Aduwa* in Hausa because it is used to make prayer boards. The Acacias can form closed woodlands, that are impossible to ride through on horseback. The dry zone mahogany, *Khaya senegalensis*, may also be found, usually near streams. The branching palm, *Hyphaene thebaica*, is common on dry well-drained sandy soils. In the low lying seasonal swamps, the black cotton soils are typical. These soils are rich in montmorillonitic clays, that result from the soil-forming processes in arid areas; they are flooded and waterlogged in the dry season, and desiccated and cracked in the dry season, often containing the burrows of the lung fish, *Protopterus annecteus*.

In the soils of the dry savannas, owing to the low rainfall, leaching is reduced and base accumulation may occur in the topsoil. Moreover, because of the effect of the long dry season preventing decay of organic matter, humus may accumulate in the soil. Hence the soils of this region tend to be fertile and very suitable for arable agriculture, particularly for crops of cotton, groundnuts and millet. However the growing season is short, lasting perhaps only two or three months. Where the water table is not too deep during the dry season, and is accessible to tree roots, the zone may be suitable for tree crop plantations, particularly of the exotic species *Azadirachta indica* (Neem) and of the Petford and Katherine provenances of *Eucalyptus camaldulensis*. On moister more southerly parts of the zone, the Cashew *Anacardium occidentale* may also be planted. There have been instances of plantations of Neem and of Eucalypts rooting into perched water-tables, and the stands dying off after a few years of good growth, when transpiration of the leaves became sufficient to exhaust the water reserves. However, the falling water tables associated with citrus plantations in parts of Israel and the southern United States are due rather to pumping irrigation water from bore holes. Trees normally assist the conservation of water, as explained later in this chapter.

The forest regions have only occasional strong winds, mostly at the time of the onset of the rainy season. These are strong spiralling gusts of wind, called **tornadoes**, which may blow trees over, or very frequently twist the crowns off them or break major branches. Only the savannas, and particularly the Sudan and Sahel savannas, are subject to strong prevailing winds, causing sand drift during the dry season. This can be a serious problem, particularly because of the desertification that has occurred due to a succession of low annual rainfalls since 1973. Farm crops, and even houses, may be smothered by sand; or the sand may be eroded away, leaving the farm crops prostrated on the surface. The vegetation of the Sudan savanna is sharply distinct from that of the Guinea savanna, because of the change in parent material from which their respective soils are derived. In the wetter more southerly parts of the Sudan savanna, the tree species *Anogeissus leiocarpus* may form a considerable proportion of the vegetation. This species is found also on sufficiently deep patches of sandy soil throughout the Guinea savanna zones.

Effects of human activities

Sudan savanna

Possibly because of its inherent fertility, the Sudan zone contains large cities, and considerable concentrations of population. The famous city of Kano has existed for at least 1000 years, and in 1952 already had a population of 131 000 supported substantially by the surrounding farmlands. Permanent agriculture was practised on the traditional town farmlands, which extended around its walls. Every morning there emerged through the city gates donkeys carrying panniers of dried **night soil** (human excreta) out to the farms, and bringing back firewood cut from farm trees, to fuel the city's cooking fires. This created a balanced ecological system; though the extent of the permanent farmlands was limited by the distance a donkey could travel in a day.

More distant from the city were bush farms, managed on a fallowing system without manuring. Typical farm trees were *Acacia nilotica* (the fruits of which are used to tan 'Moroccan' leather), *Adansonia digitata* (Baobab), *Acacia albida*, *Balanites aegyptiaca* and *Tamarindus indica*, the fruits of which are useful as condiments, food or fodder. The trees also assist in maintaining soil fertility, by bringing up nutrients from lower soil layers and redepositing them on the surface in leaf litter, to replace what is removed in the crops or washed down the soil horizons by leaching.

Protracted over-cultivation may cause the topsoil to become leached white, possibly accelerating pan formation by the deposition of iron in the subsoil. Such soils are deficient in both nutrients and micro-nutrients, and are left to be colonised by the shrubs *Guiera senegalensis* and *Calotropis procera*, which indicate poor soils. Surface erosion can expose the sheets of underlying laterised pan, and cultivation of the land is abandoned. It then becomes covered by scrubby thickets of *Combretum micranthum*. Repeated use of artificial fertilisers makes it possible to extend the years of cultivation, but the progressive loss of unreplenished humus reduces both the inherent fertility and stability of the topsoil, which in consequence begins to blow away.

Clearly the traditional methods are not designed for remunerative human employment. More sophisticated methods of sewage collection and disposal could replace it; and the internal combustion engine could be used instead of the donkey for transport, as well as for farm

Shadoof irrigation

traction and cultivation – for which the presence of the farm trees would be a nuisance. In the event of the town planners bending their minds to this matter, such innovations would not be as enduring as the traditional system, though nevertheless this is doomed under present day socio-economic pressures.

Irrigated farming, or rather gardening, has traditionally been practised in this area, using the **shadoof** or long pole, pivoted and counterweighted, suspending a skin bucket on a rope to swing water from a water hole into the irrigation channels of the garden. These gardens are used during the dry season to raise crops such as carrots or onions, which are also exported from Kano to other cities in the country.

Irrigation of farmlands has seemed an obvious way to extend the farming period, to provide more regular employment to farmers and to increase agricultural productivity. During recent years in Nigeria, the Government has invested considerable sums of money in large-scale irrigation schemes. But the general experience, and not only in Nigeria, is that they cause considerable dislocation, both to the environment and to human communities. They also represent a considerable capital investment which may not be justified by the increases in farm yields, because of the comparatively small number of people to benefit. The cost benefit to society may be less than alternative expenditure, for example by providing dry season employment, and promoting farming somewhere else better watered naturally. Wrongly employed irrigation can cause problems with salination of the topsoil, or with drainage; traditional fisheries can be ruined, and the snail-borne disease, Bilharzia, can be introduced into a new area. In Sokoto State there have been violent protests from a disillusioned populace as a result of an inadequately conceived scheme. But nevertheless it is apparent that irrigation is capable of making very significant contributions to agricultural productivity, as it does in Asia. Clearly caution is needed.

The droughts of 1973, and again of 1983, with the intervening dry years, have resulted in desertification of the Sahel zone and the more northerly parts of the Sudan zone. In 1983 millet crops failed due to inadequate rainfall and to the early termination of the rains. Cattle had to be slaughtered, or died, because of the lack of both water and fodder – a situation aggravated by the late start of the following rainy season. Affairs have been made worse by an epidemic of Rinderpest which has sometimes wiped out whole herds of cattle. Of course this is not the first time, for there were serious droughts also in the early 1930s and previously, at intervals, in the nineteenth century. Rinderpest first reached Africa in the 1880s.

It may be of interest to note that it has been the loss of their herds that has, in the past, been the main factor causing the nomadic Fulani to settle and to turn their hands to arable agriculture.[3] It may be an ill wind that blows nobody any good; this forcible fusion of pastoralist and arable cultivator may prove to be exceedingly beneficial for the development of farming in West Africa. Cattle can be exceedingly destructive when their numbers become too great. The Sudan and Sahel zones have always been important for their nomadic cattle herds, not least because they are free from the scourge of the Tsetse fly, which infects man and beast with sleeping sickness. However cattle can eat off the surface vegetation, and trample the surface of the ground making it vulnerable to erosion.

The herdsmen exacerbate the situation by lopping the trees so that the cattle may browse on the foliage. They also set fire to the dried vegetation in order to burn off the dead material and to promote a new flush of grass and tree foliage. This adds to the destruction, and leads to a progressive reduction in tree numbers. Goats complete the damage, because they are browsers rather than grazers, and will strip the bark from trees, besides preventing the growth and establishment of young tree regeneration. Firewood collection gives the *coup de grâce*, and if wood is in short supply people will even dig out the stumps and roots of the trees as fuel. In this way a well wooded area can become divested of its vegetation and exposed to the full force of sun, wind and rain.

Trees are a natural advantage because of the protection they give to the land. They check the strength of the wind and prevent sand from drifting; they check the impact of the rain by intercepting it with their foliage, and they check the force of the sun, preventing the topsoil from drying out. When the leaves are shed, they fall to the ground as leaf litter, which mulches the soil and protects it from sun, wind and rain. The leaves decay, encouraging the activity of the soil fauna, and become incorporated in the soil, returning nutrients and restoring the humus content. Moreover, the loosening of the soil due to these natural processes encourages infiltration of rain, making more water available to the roots of growing plants, and at the same time reducing surface erosion. Lower windspeeds cause a reduction in the loss of water by evapo-transpiration, so that neighbouring crops lose less water, the stomata in their leaves remain open longer so as to protract photosynthesis, and therefore crop yields are enhanced. It is stated that wind velocities are reduced detectably at a distance to leeward

3 Wilson-Haffenden, J.R. (1930) *The red men of Nigeria*, Seeley Service, reissued Cass 1967 (p. 318)

fifty times the height of the barrier, and ten times to windward.[4] The barrier is more effective if it 'leaks' because this reduces air turbulence, and therefore irregularly structured shelterbelts, consisting of mixtures of leafy tree species, perform best. The interception of rain by the tree crowns, and the improved retention of water by the soil body, as well as evaporation of water from the foliage by transpiration, reduces the tendency of soil to become waterlogged or to flood at times of heavy rain. At the same time, dry season stream flow is improved. Because the soil is kept moist, but not waterlogged, there is enhanced activity of soil fauna which improves nutrient cycling.

In the Kano area, it was hoped to establish over 5000 ha of shelterbelts before 1985, and in the Katsina area 3700 ha before 1990. The shelterbelts were to be oriented at right angles to the prevailing wind. Each belt would be twenty-seven metres wide and up to two kilometres long, made up of eight to ten rows of trees at three metre planting spacings. The belts are laid out in series of five rows with intervals of 200 metres laterally between them. The two centre rows may be of *Eucalyptus camaldulensis*, with side rows of *Azadirachta indica* or *Acacia* species or, in the moister parts of the zone, of *Anacardium occidentale*. The *Acacia* species have the disadvantage of being leafless during the dry season, with the exception of *Acacia albida* which is leafless in the rainy season. It was also proposed to manage the shelterbelts for firewood production and for building poles, on an eight year rotation. Half the width of the belt would be felled every fourth year alternately, and this is expected to give an overall average yield of three to four m^3/ha every year. This should be beneficial, as the coppicing will prevent the belts from thinning out at the bottom as the trees grow, so that this controlled exploitation will improve their effectiveness.

It is anticipated that the arable crops on the intervening land should give at least 10% increased grain yields. In fact experimental plots at Yambawa gave yields of groundnuts and millet considerably more than double those for plots outside the shelterbelts. In this densely populated area, farmers were reluctant to release land for shelterbelt planting, and the area they had handed over to the Forestry Department had deteriorated so badly under repeated cultivation that farming on it had been abandoned. However, when farmers saw the improved yields being obtained from the arable crops in the experimental plots, they returned to the area to farm themselves. It

4 Kittredge, J. (1948) *Forest influences,* McGraw-Hill (p. 394)

was reported that, during the severe drought of 1983, crop yields from the protected area were average, whereas crops in the surrounding farmlands failed to mature and were virtually a total loss. No better advertisement for the merits of shelterbelts was needed to convince a sceptical population.

Although a shelterbelt area of 30 hectares protects 200 hectares of arable land, the cost of establishing the belts is formidable: US$5000 per hectare of actual plantation. Half or more of this amount is swallowed up by the fencing needed to protect the young trees from livestock. The expense is justified only by the present dire predicament, brought about by chronic drought, and because of the substantial rural populations which depend on the land for their livelihood.[5]

In a land-use report, it is recommended that for maximum effect the width of shelterbelts should be less than five times their height, and that only trees of the same growth rate should be planted in the same belt, so that the sides of the belt are vertical. The report asserts that this is more effective than graded tree heights, which may give an aerofoil effect and hence no shelter to leeward. The belt should be permeable, and a single row or pair of rows of trees only is preferred. That is, in effect, hedges.

Guinea savannas

The Guinea savannas, on basement-complex derived soils, are by comparison much less suited for arable cropping. The landscape tends to be rolling, and the soils to be highly variable depending on their position in the catena. They are often stony or gravelly, with underlying layers of indurated plinthite, and are vulnerable to erosion by water. Surface streams may dry up in the dry season, and it may be difficult to reach down to the permanent water table. The Guinea savannas tend to be among the least populated vegetation zones of West Africa, and are often sparsely populated by the least sophisticated tribes. Pockets of soil can be found that are suitable for arable cultivation on a peasant scale. For extensive mechanised cultivation, heavy equipment is needed, and because this involves

5 Estimates suggest that the value of the wood yield from shelterbelts falls far short of the cost of producing it; but a ten per cent increase in arable crop production would cover the establishment cost of the shelterbelts, and a thirty per cent increase would give a **financial rate of return** (FRR) of 4.8 per cent (i.e. a rate of interest on capital invested, such that expenditure exactly equals receipts).

removing all the trees and destumping, the land becomes very vulnerable to damaging erosion. Devastation due to attempted large-scale mechanised farming may, for example, be seen at Mokwa, near Jebba in Nigeria. If agriculturists forge ahead with unsuitable methods of farming, they may leave extensive derelict landscapes that could endure for a longer time than one cares to contemplate. Nevertheless, Guinea savanna soils are perhaps more suited for arable cropping than the forest soils to the southward. They are capable of good yields of cereals such as maize and sorghum, averaging over three tonnes per hectare. Because of the lower rainfall and the existence of soils derived from igneous rocks throughout the area, they are generally less vulnerable to the effects of leaching.

For the forester, the Guinea savannas presented severe problems for forest development. Savanna woodland, when clear felled, yields about seventy cubic metres of solid wood per hectare, mostly suited only for firewood. Though most of the trees will coppice, the woodland does not recover its original stock level even forty years after initial felling. Productive woodland can only be established by afforesting with tree species that will grow rapidly and can close canopy, shading out the grass. Before 1960, the shallow indurated soils of the Guinea savanna were an intractable problem. Unlike arable crops, trees have to survive the dry season.

At this time, Eucalyptus species succumbed to termite attack, and Pines did not survive long after planting out in the field. However, during the 1960s, ways were found of establishing plantations of Eucalypts and Pines. For Eucalypts it was discovered that the addition of one gram of two per cent Dieldrin agricultural dust to the potting mixture of each plant would protect it throughout life; for Pines, the correct mycorrhizal fungus must be inoculated into the potting mixture, because these fungi are not present in the soils of the African lowland tropics. The roots of *Eucalyptus camaldulensis* were found to be capable of penetrating indurated layers of soil. A widespread dieback of some Eucalypt species on certain soils was identified as being due to a deficiency of the micro-nutrient (or trace element) boron, and was prevented by an application of 120 g of borax near the base of each tree soon after planting, together with a similar quantity of NPK fertiliser. From soil surveys it is estimated that about half the area of the savanna forest reserves is suitable for planting with Eucalypts or Pines. The fact that the other half is still not suitable perhaps underlines the difficulties presented by Guinea savanna soils.

Extensive tracts of the Guinea savannas have fallen under the sway of the Tsetse fly. This viviparous fly spreads the deadly sleeping

Zebu and Ndama cattle

sickness (Trypanosomiasis) to man and beast. Cattle are particularly vulnerable in the savanna zone as, during the heat of the day, both cows and flies seek the shade of scattered trees, and are brought into fatal proximity. The humped Zebu cattle of the nomadic herdsman are particularly susceptible to Trypanosomiasis, though the native short horn cattle (Muturu or Ndama) are resistant, and herds of the latter were formerly widespread through both the forest zone and the non-Muslim areas of the savanna of Nigeria. However the Muturu seem to have become much less common during recent years, perhaps as the transport of cows from the nomadic herds of the north has been able increasingly to satisfy the meat markets. Some protection is given to the cattle by an injection of a drug, but reservoirs of infection remain in antelopes and other game.

In the 1950s an attempt was made in Nigeria to prevent the continued spread of the fly by cutting a tree-free zone across the country, the Anchau corridor, which it was hoped the fly would be unable to cross. So far as I am aware, this expectation was not realised. During the 1950s I saw human populations in the Biu area decimated by the fly. Oddly enough it was the men who were dying and not the women, so that one came across villages peopled almost entirely by women and children. Perhaps the men were more exposed to being bitten, because they went to the fields to farm, possibly removing their shirts before exerting themselves.

In the more densely populated areas, where most of the bush has been cleared for farming, the incidence of the fly diminishes. Nevertheless a consequence is that cattlemen and foresters tend to have opposing views: the foresters wanting to conserve trees and the cattlemen to eliminate them!

High forest zones

The other belt of human occupation is along the northern fringe of the high forest, where it abuts on the savanna. A reason is that base accumulation also occurs in the soil under the drier forests. These soils may contain significant amounts of montmorillonite, and are more fertile than either those of the southern Guinea savanna or those in forests under heavier rainfall. An added advantage is their accessibility to human communication via the open savanna country of the hinterland, permitting the evolution of organised government. In West Africa, these forests are frequently settled by city dwelling peoples from the savanna zone. This belt is characteristically under-planted with cocoa, particularly where the soils are derived from granites of the basement-complex.

The more leached soils to the southward are planted with orchards of *Cola* species, from which the nuts are exported to the north. They are also suitable for planting robusta coffee. The topography under these drier high forests is usually rolling, resembling that of savanna. Rocky inselbergs may jut above the surrounding forest, indicating periods of drier climate and savanna-isation during the recent geological past.

There was also a strip of human settlements along the coast. The people supported themselves by fishing, and could communicate with each other by boat via the creeks, the lagoons and the open sea. Between the coast and the forest/savanna interface were often extensive tracts of unpopulated, or sparsely populated, forest which cut off the coast from the hinterland. These could be crossed only by intrepid travellers, who risked starvation before reaching the other side. The only easy access was where navigable waterways facilitated the journey from the coast to the savanna country. In this forest zone, travelling southwards towards the equator, the soil colour changes progressively from predominantly reddish to predominantly yellowish. This is because, under conditions of high ambient temperature and heavy rainfall (say exceeding 2000 mm per year) the soils become deeply weathered. There is more marked decomposition of the mineral soil fraction and greater removal by leaching of the decomposition products, such as exchangeable bases and silica. These soils have a low base saturation, and may be extremely acid, down to around pH 4, with the iron held in the ferric form. The soil becomes substantially composed of kaolinite clay (i.e. aluminium silicate) and sesqui-oxides of aluminium. Free aluminium may be present in the soil. Due to the higher rainfall, the soils are more leached and the terrain becomes deeply dissected.

On loose sandy deposits on steep slopes, I have observed surface erosion even under primary undisturbed forest (in Cameroon and Sarawak). These soils are unsuited to arable agriculture, because of the difficulties of maintaining fertility. However, they have proved very suitable for commercial plantations of oil palm and para rubber. In a warm wet climate, decay is rapid and tropical forest maintains itself on a relatively low capital of fertility, which is quickly recycled from the soil back into the trees. The forest biomass is typically between 150 tonnes and 350 tonnes per hectare, whereas in temperate forests it may exceed 500 tonnes per hectare. In equilibrium, decay and growth are counterbalanced; trees of climax forest tend to be relatively slow growing, and biological productivity is low compared with timber crop plantations. The range of species may exceed fifty within a single hectare, and the size classes range up to well over 150 cm bole diameter. This adds to the difficulty of exploiting natural forest, which does not pay for its own removal. Consequently, when forest is replanted with tree crop plantations, or cleared for peasant farming, the greater proportion of the wood is burned on the site.

Throughout savanna and forest zones, there may be patches of soils derived from volcanic parent material, which because of their higher inherent fertility have attracted human settlements. These soils are rich in montmorillonitic clay and in exchangeable bases. They retain their productivity even under repeated and protracted cultivation. At Ikom in Nigeria, near the Cameroon border, despite the high rainfall it is possible to grow cocoa, or to reforest with Teak plantations. Around Cameroon mountain there are concentrations of population on soils derived from old lava flows, whereas the surrounding sedimentary soils are more or less bereft of human settlement. Dense populations occur on the basaltic plateaus of the Cameroon highlands. Dotted across the lowland savannas of West Africa, are local foci of population, wherever volcanic intrusions have occurred.

Natural high forest

Montane tropics

The montane and high altitude plateaus of tropical Africa, above 1200 metres elevation, are not within the scope of this book. The reasons are that, in the cooler atmosphere at high levels, soil conditions resemble more those of temperate climates, and the techniques of temperate agriculture can be applied almost directly. At these altitudes, many naturally occurring plant genera are immediately recognisable to a European botanist: for example, *Alchemilla*, *Anagallis*, *Brachypodium*, *Carduus*, *Clematis*, *Deschampsia*, *Delphinium*, *Festuca*, *Geranium*, *Gladiolus*, *Hypericum*, *Impatiens*, *Juncus*, *Lobelia*, *Luzula*, *Plantago*, *Poa*, *Ranunculus*, *Rubus*, *Senecio*, *Sonchus*, *Trifolium*, *Verbascum* and *Veronica*. The fern, Bracken (*Pteridium aquilinum*), may be widespread.

The soils are often derived from volcanic rocks, and consequently are fertile. During my stay in Cameroon in the 1970s, the farmers of the highlands were acclaimed as paragons of industriousness and productivity. It was recommended that the Agricultural College should be moved from Yaoundé to Dschang at 1500 metres altitude, so that students could see an example of how agriculture should be done. Although the highland tropics are extremely pleasant places in which to live, especially for Europeans, the lessons would be irrevelant to the lowland humid regions that cover the major part of the country, and could have been counterproductive to national needs. It is in the lowland tropics that the problems of agricultural development particularly lie, and about which this book is written.

Effects of farming

In the humid lowland tropics, organic matter breaks down relatively rapidly. When forest is cleared to make way for arable crops, half of the organic matter may be lost from the soil after about five years. This has damaging consequences for the soil, which loses its ability both to hold nutrients and to retain its structure. Even in the second year of cultivation, the soil noticeably lacks tilth compared with the previous year, and shows a 'flat' consolidated appearance. Moreover, many of the bases released by burning the original forest have been leached from the topsoil. Some crops, such as plantains and yams, may be very sensitive to the loss of organic matter; they may not grow satisfactorily except when they are planted during the initial year. After two or three years' cultivation, crop production may have deteriorated to such a degree that recropping is not worthwhile, because yields may have fallen to half or less of those of the first crop. Additions of chemical fertiliser may give diminishing returns, due to the progressive depletion of humus, and therefore the loss of the ability of the soil to hold nutrients against leaching. In areas of Africa with low rainfall, leaching may still be significant because the precipitation occurs during two or three months only, and during this period rainfall exceeds evapo-transpiration.

Besides loss of fertility due to leaching, the soil is also vulnerable to erosion. This worsens as the diminution of soil organic matter with time reduces the ability of the soil to retain cohesion and structure. During heavy rain the fine particles of the soil may be washed out of the surface soil, eventually leaving a soil covering composed mainly of gravel and stones. During cultivation the soil becomes consolidated, and a substantial proportion of the rainfall fails to infiltrate the soil, but adds to the surface flow, streaming away as brown muddy water to swell the streams and rivers.

The muddy colour of the rivers is the first impression made on a stranger from temperate climes, visiting the tropics for the first time. But this is not the end of the damage that may be caused. The complete topsoil may be stripped away, leaving the subsoil exposed, virtually destroying the productive capacity of the land. Gulleying may occur on slopes, biting into the landscape, and sometimes carrying away roads and houses in the process. This is not the finish: due to the burden of earth in the water, streams and rivers become silted up and prone to serious flooding. Moreover, the increased surface flow and reduced infiltration, causes stream flow to become more irregular and less sustained; thus, a stream floods at times of heavy rainfall, and completely dries up during periods of drought. At one period crops may be washed out of the ground and livestock drowned, and at the other they may perish for want of water and fodder. Lakes and reservoirs themselves become silted up and useless. In the Philippines I saw a hydro-electric reservoir, for which the original scheduled life was 60 years, where silting up would occur in 30 years, owing to the activities of peasant farmers returning to areas that they had been compensated to leave – perhaps plunging a whole region of the country into darkness and damaging its prospects of economic development.

Peasant cultivation

The peasant farmer prepares land for cultivation during the dry season, by cutting the undergrowth and then felling the trees. He allows this material to dry for a month or two, before he sets fire to the farm. After the burn, he cuts and stacks the debris into bonfires, and he may place these round the base of trees that are too big or too hardwooded to fell, so that the tree is killed by the fire. The ash returns the fertility to the soil, and the soluble nutrients are held mainly by the organic matter (humus) that is contained in the soil. Tilth and crumb structure are good, because of the binding properties of the humus. The quantity of organic matter is small compared with that in temperate soils, but the amount is highly correlated with the fertility of the soil. The heating effect of the fire is also beneficial, because it kills off snails, insects, smaller mammals and other small vermin, besides enhancing nitrogen content due to the semi-sterilisation of the soil. In parts of the savannas of Central and East Africa, the **citemene** technique is practised, in which wood is cut from a more extensive tract of woodland and carried onto the plot to be cultivated, where it is burned. This concentrates fertility on the patch to be farmed, at the expense of a much larger area. I visited an experiment near Kaduna, where four different treatments were tried, each repeated several times to give confidence in the result. In the first treatment, the vegetation was cut and removed unburned before cultivation; in the second, the vegetation was cut and burned on site; in the third, the vegetation was cut and burned but the ash was removed afterwards; in the fourth, the vegetation was cut and removed without burning, but the ash from the third treatment was placed on these plots. Subsequently the plots were sown with a species of Eucalyptus (*E. citriodora*). The results after a year showed that growth was best for the second treatment, followed by the third treatment, and clearly demonstrated that the heating effect of the fire on the soil had an important effect on soil fertility, besides the mere addition of ash. The worst growth was given by the first treatment, in which no burning occurred.

The farmer scrapes the topsoil into mounds or ridges. This is not only for drainage and aeration of the soil, but in order to make the topsoil deep enough for his root crops, such as yams or cassava. He does not usually attempt to sustain or improve soil fertility using household refuse, ashes, compost or human or animal manure. Indeed his fields may be beyond a distance over which these could be carried easily. As the fields are unfenced, they must be far enough away to avoid damage by free-ranging village goats, sheep, pigs or poultry. Moreover the use of excreta as manure may be anathema to him. As the population increases, the pressure on the land intensifies; and as the land is refarmed more frequently, the fallow period becomes shorter and shorter. If land is rested for seven years it may,

Cultivation mounds used by peasant farmers

Palm bush

perhaps, have half the fertility of newly cleared forest land; but the fallow may become reduced to as little as two or three years, with less than a quarter of the crop yields from new land. The friability of the soil worsens perceptibly, and the peasants may have difficulty growing certain important food crops, for example plantains or yams. Plantains may become confined to the borrowpits amongst the houses, into which the villagers have thrown their rubbish.

Two important processes occur under arable cultivation: the nutrients are leached progressively from the soil by rain, and the organic matter decays and diminishes in amount. The latter is crucial, because it reduces the ability of the soil to hold nutrients and to retain crumb structure and tilth. Even additions of fertiliser may become unable to maintain soil fertility. On lighter soils the effects are noticeable more rapidly, and after two years farmers may be unwilling to continue cropping. If the land is cultivated for periods exceeding two years, then more and more woody plants are killed out during each cycle of clearing and farming, and there is a progressive change to grassy fallows. These are much less effective at restoring soil fertility than woody fallows, especially where the grass is Spear grass (*Imperata cylindrica*). If the land is kept open, soil degradation proceeds to a degree where trees have difficulty re-establishing themselves. Ultimately it may become difficult and expensive to rehabilitate the land.

The African peasant practises a system of shifting cultivation, or **bush-fallow** rotation. The stumps and roots remain in the ground after the farm is cleared for cultivation, stabilising the soil and rotting progressively to release nutrients. When the farmer abandons cultivation, the surviving stumps produce coppice shoots and root suckers, which quickly re-establish a vegetation cover. After about 15 years fallowing the soil may be restored virtually to its original levels of fertility; but it may take 50 years before the tree growth begins to resemble mature forest. The humus content of the soil, its fertility and the activity of the soil fauna, all increase so that the soil structure is restored. However Nye and Greenland write, 'The traditional system of shifting cultivation in the forest admirably protects the soil from erosion in spite of steep slopes and heavy rainfall. In savanna, erosion losses during a single period of cropping are usually slight because of the gentle slopes, and even on steeper slopes they are not sufficient to terminate the immediate cropping period. Repeated cycles of cropping and fallowing may however cause total ruin to savanna soils because they uncover subsoils rich in iron oxide, which if not indurated already, become so on exposure.'[6]

Trees useful for their fruits, or for other purposes, were retained in the farm. In savanna zones, farm trees include *Adansonia digitata*, *Acacia albida*, *Butyrospermum paradoxum*, *Parkia biglobosa*, *Tamarindus indica* and *Vitex doniana*. In forest zones trees preserved include *Chrysophyllum albidum*, *Cola acuminata*, *Dacryodes edulis*, *Irvingia gabonensis*, *Pentaclethra macrophylla*, *Treculia africana* and the oil palm *Elaeis guineensis*. In high rainfall areas that are densely populated, the natural vegetation may become converted to **palm bush**: oil palms forming a semi-open woodland, with cultivated patches and fallow beneath. Arable crops include coco-yam, which is relatively shade tolerant. The palms are a very useful crop, and provide roofing material and reinforcing in mud walls, edible oil, soap, illumination and fire lighters.

Mechanical cultivation

Perversely, mechanical cultivation requires the removal of stumps and roots, and the debris is pushed into 'windrows' for burning.

6 Nye, P.H. and Greenland, D.J. (1960) *The soil under shifting cultivation*, Technical Communication No 51, Commonwealth Agricultural Bureaux (p. 156)

Moreover, heavy machinery compacts the soil, impeding drainage and root penetration; and because tropical topsoils are naturally only about 20 cm deep, they are easily scraped away by the action of bulldozer blades, or by slippage of the tracks. More recently the knife-ground (KG) blade has been introduced, which shears off woody growth flush with the ground.

Large boles may be split with a 'stinger' at the side of the blade, until they can be sheared off; but ploughing is impossible without root removal. After windrowing and burning, no-tillage methods are used in which weed growth, and the crops themselves after harvesting, are killed off with herbicidal sprays. At Ibadan, Lal found that traditional peasant farming caused erosion of 0.02 tonnes per hectare during the three years following deforestation, whereas mechanical land clearing with tree pusher and root rake followed by conventional tillage gave a soil loss of 24.3 tonnes per hectare over the same period. Even the least damaging mechanical method, employing shear-blade and no-tillage, caused erosion of 4.8 tonnes per hectare (see *Table 5*). In fact, erosion following conventional tillage was so severe that, despite contour ridging, cultivation had to be abandoned prematurely and a cover crop of *Mucuna* sown to cover the soil quickly and to stop further soil loss. Yet Roose comments that development programmes in West Africa have been less concerned with increasing yields of existing farms, than with extending cultivated areas by mechanisation of land clearing and sometimes of entire cropping operations. Lal points out that topsoil forms at a rate of about one centimetre per decade, so that it could take well over a century for denuded topsoil to become restored.[7]

Hartmans writes, 'In some places where inappropriate land clearing and subsequent soil management methods have been used, erosion is removing annually 50 tonnes or more of the surface soil per hectare. This, incidentally, is more than 100 times the natural rate of soil formation in the tropics, even under ideal conditions. Failure to introduce effective soil conservation measures will by the end of the century result in an approximately 16 per cent loss in the area of rainfed cropland, and a 30 per cent loss in the productivity of the land. IITA [International Institute of Tropical Agriculture] scientists learned the hard way . . . Within five years of constructing and laying out the Institute's research farm according to the most scientifically

7 Papers by Lal, R. and Roose, E.J. at a symposium on *Land clearing and development in relation to environmental protection in the humid and subhumid tropics* at the International Institute of Tropical Agriculture, Ibadan, November 1982

Table 5 The effects of deforestation and post-clearing soil management on run-off and erosion from an alfisol for maize/cassave/maize/cowpea rotation from 1979 to 1981 (land was cleared in 1979)

Treatment area	**Basin area** (ha)	**Runoff** (mm) *1979*	*1979–81*	**Soil erosion** (t ha^{-1}) *1979*	*1979–81*
Forest	15.0		*T*	*T*	*T*
Traditional farming	2.6	3.0	6.6	0.01	0.02
Manual clearing/ no-tillage	3.1	16.0	16.1	0.4	0.4
Manual clearing/ conventional tillage	3.2	54.0	79.7	5.0	9.8
Shear blade clearing/ no-tillage	2.7	86.0	104.8	4.0	4.8
Tree pusher, root rake/ no-tillage	3.2	153.0	170.0	15.0	15.7
Tree pusher, root rake/ conventional tillage	4.0	250.0	330.6	20.0	24.3

From Lal, R. (1983) *Soil erosion in the tropics with particular reference to agricultural development and soil management,* proceedings of symposium on *Hydrology of Humid Tropical Regions with Particular Reference to the Hydrological Effects of Agriculture and Forestry Practice,* Hamburg August 1983, IAHS Publication no. 140 (pp. 221–239)
Note: *T* = unmeasurable trace

sound Western technology, the land was deteriorating fast and crop yields were declining. As a result, part of IITA's research farmland is almost completely destroyed. The fancy terraces and contour bunds didn't work out as expected on the fragile tropical soils. IITA was faced with a visibly embarrassing example of how not to farm in the tropics. Now, of course, we know that when the bush is cleared for cultivation, the soils change for the worse rapidly and irreversibly. The soil becomes less permeable and more vulnerable to erosion.'[8]

8 Hartmans, E.H. (1984) *Prospects for Nigerian Agriculture*, Distinguished Lecture Series, Nigerian Institute of Social and Economic Research, University of Ibadan, Ibadan, December 1984

Taungya farming

A palm plantation

A rubber plantation

Tree crop cultivation

Ahn writes, 'There remains the question of how to reduce or prevent yields falling with successive cropping, or how to restore productivity by means other than fallow when they do fall. No continuous system of arable mechanised cultivation has yet proved workable and economic in the conditions which prevail in the forest zone of West Africa, where the most profitable and best understood type of large-scale operation is still often the planting of tree crops.'[9] Indeed humus levels can be difficult to maintain in the soil without the intervention of a tree phase, and peasant farmers have also turned to trees as cash crops: cocoa, coffee, cola, oil palm and rubber.

In West Africa, foresters have had long experience of cultivating arable crops to assist in the establishment of timber crops. This system is known as *taungya* – a Burmese word for 'hill farming', i.e. shifting cultivation. This is usually in collaboration with peasant farmers, who are glad to obtain access to virgin forest soils to cultivate, and to retain their own lands under perennial cash crops. The forester gets forest land cleared and cultivated at little or no cost, and plants his trees among the arable crops of the farmers. Meanwhile the farmers practise their customary mixed cropping with little constraint, though they may be discouraged from planting cassava at the beginning of the farming cycle, and only allowed to put plantains at the edges of their plots, to avoid smothering the young trees. The traditional Ibo varieties of cassava are preferred, because they are straight-stemmed and have been selected to be suitable for intercropping, though they may be less vigorous than the traditional heavy branching varieties. Normally farming continues until the trees close canopy after two to four years. The farming cycles used at various locations are given in *Table 6* opposite.

More recently, because of the need to increase the area of timber plantations, at a time when numbers of willing peasant farmers are becoming smaller due to the exodus from the land, foresters have themselves begun to raise arable crops of maize, cassava or groundnuts. Combinations of arable crops and timber crop plantations are also being termed **agro-silviculture** or **agro-forestry**.

Most trees experience considerable improvement in growth because of the arable cultivation, which is more intensive than could be afforded without the arable crop. In fact it may be essential for satisfactory growth of Teak (*Tectona grandis*), a species best suited to

9 Ahn, P.M. (1970) *West African soils*, Oxford University Press (p. 252)

Table 6 Farming cycle, taungya farms

	West (Gambari)	**Midwest (Sapoba)**	**South-East (Awi)**
Mean annual rainfall (mm)	1250–1600	1500–2000	1700–2600
Geology	Basement complex	Sedimentary sand	Basement complex
Farming agency	Peasant farmer	Peasant farmer	Forestry Dept
Year 0			
October	–	–	Demarcation
November	Demarcate farms	Demarcate farms	Brushing Felling
December	Brushing		
Year 1			
January		Brushing	Burning & packing
February	Felling		
March	Burning and packing	Felling Burning and packing	Sow maize (1 × 1m)
April	Sow maize, Egusi melon, okra Dig yam mounds	Sow maize, melon, okra, amaranthus and peppers	–
May	Plant yams when maize germinated	Dig mounds and plant yams	–
June	Harvest maize	–	Harvest maize
July	Harvest melon	Harvest maize and melon	Plant cassava
August	Plant cassava		
September	Resow maize	–	Resow maize
October	–	–	–
November	Harvest 2nd crop maize	Harvest yams	Harvest 2nd crop maize
December	Harvest yams		
Year 2			
January	–	Plant cassava	–
April–August	Harvest cassava	–	–
July–December	–	–	Harvest cassava
Year 3			
January–March	–	Harvest cassava	–
The tree crop is planted in Year 1 during May and June			

A timber crop plantation

the base rich soils of the drier forest. The arable crops benefit from the residual fertility of the forest, which would otherwise leach uselessly away. On sandy soils under higher rainfall, the system is used to establish mixtures of Mahoganies and Opepe (*Nauclea diderrichii*). The Mahoganies, such as *Entandrophragma cylindricum, Khaya ivorensis* and *Lovoa trichilioides*, cannot be grown satisfactorily in pure crops because of crippling attack by the insect shootborer, *Hypsipyla*, and also, perhaps, young trees of these species have insufficiently spreading crowns to shade out weeds. The *Nauclea* can be cut for transmission poles at about fifteen years, and will coppice, whereas the Mahoganies take at least sixty years to reach timber size.

On basement-complex derived soils, too leached for Teak but insufficiently watered for Nauclea (which also becomes affected by a shootborer), *Gmelina arborea* grows well. Peasants do not like this tree as a taungya species, because of its rapid growth, which smothers their crops. At a normal planting spacing of 2½ or 3 metres, it closes canopy in between twelve and eighteen months. However it is possible to obtain a crop of maize and sometimes of cassava. Because *Gmelina* is exceptionally fast growing, it is being planted for pulpwood on a rotation of eight years, for paper manufacture. Standards can be retained amongst the coppice, to grow to timber on a sixteen year rotation.[10] Other timber tree species suitable for taungya in West Africa are Obeche (*Triplochiton scleroxylon*), Idigbo (*Terminalia ivorensis*) and Afara (*Terminalia superba*). Though these species are more common naturally in drier forest, they grow best when planted on higher rainfall sites. They provide a relatively light canopy, and are highly suitable as overhead shade for cocoa and coffee.

Taungya may also be practised on savanna soils; for example in the Sudan zone, Neem (*Azadirachta indica*) may be established among an arable crop of groundnuts. However for Eucalypts the practice of arable cultivation may give more difficulties. In West Africa the *Eucalyptus* species are extremely sensitive to termite attack, and the presence of an agricultural crop may encourage termites. The dose of insecticide applied in the seeding potting mixture protects the root collar from termite attack, but the cultural operations may push soil around the stem of the young tree, furnishing a bridge for the termites to attack it. Pines have rather slow early growth and are strongly light demanding, and may be smothered by a tall growth of corn stalks, and particularly when the stalks are thrown down after harvesting. In the Guinea savannas the competition for water by the arable crop may also detract from the growth of the trees.

Control of pests and weeds

The Forestry Department acts as 'Chief farmer', measuring out and distributing the plots among the farmers. The Department ensures that all the farmers clear their land in time and have it ready for burning, and may help the farmers by felling some of the larger trees with power saws. From the point of view solely of growth of their arable crops, the farmers prefer to have as few trees remaining as possible. It is essential that all the plots are ready for burning at the same time. A premature burn, before the woody material is sufficiently dried, or a burn that is too late so that frequent rains make it too wet, can enormously increase the work of cutting and stacking the unburned debris into bonfires.

There is an advantage in having the farms close together, because it reduces the risk from depredations by wild animals, and the intense

10 Recent research at the University of Ibadan has shown that litter fall from a ten year old plantation of *Gmelina* was almost the same as that from natural high forest, and significantly greater than that from a thirty-year old secondary forest.

fire destroys many of the small rodents and other vermin; however, the traditional peasant practice of scattered plots and mixed cropping may help to reduce the incidence of insect pests and other diseases on the various crops. Having extensive areas of the same crop can cause a spectacular build up of infestations, or of populations of crop predators. I have seen immense flocks of weaver birds (*Quelea quelea*) which darkened the sky in the evening as they flew from the ripening fields of millet to their roosting sites. Populations of the variegated grasshopper (*Zonocerus variegatus*) have burgeoned from almost nothing, as the growing of cassava has become more widespread, because it provides a refuge and a source of food for the insect during the dry season. A fall in soil fertility may also increase the vulnerability of a crop to attack by pests, for example of yams by the Yam beetle (*Heteroligus*) and of plantain by the stem borer (*Cosmopolites*). The growing of the same crop repeatedly on the same ground may also encourage its pests and diseases. Examples are spider mite and mealybug attack on cassava, soil nematodes on tomatoes and plantains, rusts on cereals and shootborer on maize. Shootborers can be particularly difficult to deal with, even with insecticides, because the insect is protected inside the plant, except against systemic insecticides which penetrate the tissues of the plant. Another example is the build up of populations of the root parasite *Striga hermontheca* (a flowering plant) on Sorghum, which can become a considerable nuisance. As the soil becomes more leached, mineral deficiencies can also become pronounced and begin to affect adversely the growth of crops.

Mature forest does not contain the weeds of arable cultivation, and the ground flora of the forest is not adapted to exposure to full sun. Consequently the farm is spared serious weed competition during the initial months of cultivation, particularly if there has been a hot burn; but weed populations build up as farming continues. The farmer may try to reduce their incidence, by hoeing the farm regularly before the weeds have time to flower and produce seeds. A negligent farmer who does not weed his farm properly may make life difficult for his neighbours. The thoroughness of cultivation has been found to have an important effect of the growth of crops, possibly even more important than that on fertiliser additions, though the effect of the two operations is to some extent additive. A farmer may also deliberately use weeds to provide ground cover during intervals between crops, such as a first and second crop of maize, hoeing them up when he introduces the second crop. Protection to the surface of the ground to prevent surface erosion is extremely important, and in some areas Egusi melon is sown and rapidly covers the ground between the young maize; the melons ripen in two or three months.

These covering crops also reduce the incidence of weeds, by preventing them from seeding and growing. However weeds can eventually become such a nuisance that they discourage further cultivation, and the farmer prefers to clear a new piece of land. Some crops give dense shade and clean the land of weeds. Examples are cassava, and potatoes in temperate climates, though nematodes can build up under a succession of potato crops, and wireworm (Cranefly larvae) under grass. The incidence of weeds, pests and diseases can be controlled by a judicious rotation of crops. However, in shifting cultivation, the farmer relies mainly on the fallow to eliminate weeds. The cultivation period is relatively short, and cassava is often the last crop in the farming cycle during the second or third year.

The situation for savanna is similar, the natural vegetation normally being different from the weeds of cultivation, annual plants tending to replace perennials. However spear grass (*Imperata cylindrica*), like Twitch or Couch grass (*Agropyron repens*) in Europe, can become a serious nuisance because of its underground rhizomes. The grass is perennial, and almost pure populations of this species may result from repeated shallow cultivations. This weed is very difficult to control by simple hoeing, and seems to have a strong power to inhibit the growth of crops or trees, perhaps by some effect on soil nitrification. The recently introduced weed Eupatorium (*Chromolaena odorata*) is fortunately able to invade, smother and shade out *Imperata* grass. Since first being noticed at Enugu in the 1940s, it has spread widely throughout the southern part of Nigeria. Eupatorium appears to be a soil improver, promoting earthworm activity and intense wormcasting; because it is superficial rooting the farmers find it easy to uproot. Unfortunately the foliage is unpalatable to livestock and the plant has no known use.

Effects of hunting

During the dry season the peasant is freed from farmwork to follow other avocations, and for some men this is hunting. Not everyone can make a good hunter. These men must be physically active, with acute sight, and with the instinctive ability to know their location and to judge direction accurately – for they do not carry compasses! They can do this even where visibility is restricted by tree growth, or where the landscape lacks pronounced features, or the sky is clouded over. They have learned to perceive animals from small visual clues. They may hunt alone or in groups. Originally anyone could hunt or fish anywhere within the territory of his community, although the professional hunter would be regulated by customary observance and ritual, to prepare himself mentally and physically for the hunt. Other people would probably be afraid to enter the forest at night. Hunters, armed with guns, use lights to reveal and dazzle the game (from the reflection of the eyes), besides catching animals with snares and spring traps. The Chief and elders would expect to be presented with part of the larger animals, such as the haunch of a buffalo, or the trunk and one tusk of an elephant, or the skin of certain animals such as lion or leopard. The hunter usually preserves the meat at his distant camp, by smoking it over a fire. Surveys in rural towns and villages in Nigeria in 1965 showed that half of the meat eaten was bush-meat. The growth of urban communities, and of personal wealth, has expanded the demand for bush-meat, which is regarded as a delicacy. A consequence has been an increase in the numbers of stranger hunters and fisherman. As they do not belong to the local community, they have no interest in conservation, and may descend on an area in gangs, virtually wiping out local populations of wild game or fish. The only animals and birds that seem to increase in numbers are the pests of agricultural crops such as bushfowl (Francolin) or the giant African rat, because of the extending area under arable cultivation.

Effects of fire

Fire has been the most important tool with which the African has attempted to control his environment. He has felled and burnt the trees that stood on the patch he cultivated, both to prevent them from competing with his agricultural crops, and also to return the nutrients stored in their boles, branches and foliage to the soil. When the trees are burnt, part of the nitrogen and sulphur they contain escapes to the atmosphere, and is lost to the crops; but the heating of the surface soil helps to promote nitrogen fixation in the soil, supplementing what is already held in the soil organic matter. The African peasant also uses fire to help in hunting and to improve the pasture for livestock. Dried grass is burnt off, sometimes in order to drive game, but also to give an unimpeded view and passage through the bush. Animals and birds must seek shelter in the remaining unburnt clumps of vegetation, and therefore become more vulnerable.

Dried grass is very poor feed, and the fire removes this material and stimulates the sprouting of new grass shoots which are rich in protein

Revitalised savanna, after early burning

Savanna after late burning

and highly nutritious. The passage of the fire also stimulates savanna trees to produce fresh foliage on which animals may browse – and this is a richer and more consistent source of protein than grass. Thus wild animals also benefit from the activities of hunters. Graziers set fire to the bush for the same reasons, in order to improve the feed available to their cattle and other livestock. The carrying capacity of natural savannas ranges between two hectares and ten hectares for every full grown beast. Overgrazing by cattle causes selection against the preferred grasses, and ultimately only relatively unpalatable species may remain. Trampling, and the more destructive grazing of sheep, can cause bare ground to appear and lead to surface erosion of the soil. Slopes often become horizontally ridged, due to the tendency of cattle to keep passing along the same tracks. However the addition of dung is beneficial to soil fertility, and the peasant cultivator often encourages the pastoralist to corral or graze cattle on his fallow fields.

The forester has been opposed to the use of fire, and has favoured protection of the savanna from fire as being the most beneficial for regeneration and growth of trees. In such circumstances, savanna can form a semi-closed woodland, with an underthicket of shrubs and climbers, and clumps of grasses and herbs. The ability to form closed woodland is greater on loose sandy deposits, where tree roots have access to the water table, than on the more variable **lateritic** soils of the basement-complex. Formerly it was thought that savanna was a fire climax, and that with complete fire protection the grass would be shaded out and closed woodland would be formed, albeit more stunted than high forest. This view appears too simplistic; on basement-complex derived soils the tree populations become more dense with fire protection, but the grass is never entirely shaded out. Savanna trees have thick bark and are resistant to fire. In the early stages of establishing the young tree, growth is often concentrated on the development of a swollen rootstock, before the trunk is formed, and only leafy shoots are produced above the soil. This helps tender growth to survive fire, but consequently early development is slow, and makes most natural savanna species unsuitable as plantation trees. Savanna trees mostly lose their leaves during the dry season, and the effect of a ground fire is to cause remaining leaves to fall off, and to stimulate the flushing of new leaves. Otherwise, without fire, the leaves flush just before the onset of the rainy season.

A consequence of a succession of late burns, that is towards the end of the dry season when the ground vegetation is thoroughly dried, is fierce fires that are capable of killing the trees and preventing tree regeneration (i.e. establishment of tree seedlings). These fires lead to degradation of the woodland, and to reduction in soil humus and ultimately to soil deterioration and erosion. In fact, attempts at complete protection have been found to increase the risk of intermittent uncontrollable late fires that do more damage than the intervening years of successful protection do good. The forester's compromise is to practise early burning, that is setting fire as early in the dry season as a burn can be achieved, when the fires are much less fierce and do little damage to trees or soil. Patches of unburnt vegetation remain in the moister patches or under the shade of clusters of trees. This provides shelter in which tree seedlings may have a chance to survive. However the more trees the less grass, and the cattleman may be happy to have a reduction in the tree density. In Sudan savanna, the sparser ground vegetation gives much less severe fires than in Guinea savanna, and on farm fallows there may be insufficient vegetation to burn at all. This is a result of the sandy soil and the shorter growing season.

The undergrowth of closed high forest is not normally inflammable; but the combination of farming and fire gnaws away at the forest margins. At the interface with the savanna, the higher rainfall and longer growing season results in a lush growth of grass, which dries out to provide large amounts of combustible material. The conflagrations kill remaining trees that have survived the forest destruction, and prevent their regeneration. As savanna species invade, the landscape alters to **derived savanna**. A total cessation of farming may permit a gradual re-encroachment of forest into the area. The aggressiveness with which this occurs depends on the extent to which the soil has deteriorated since the removal of the forest. Freely draining sands may become extremely leached following removal of the forest cover, and being edaphically dry they promote fierce fires. Shallow soils may also discourage reinvasion by forest. The shrubby weed, Eupatorium (*Chromolaena*), may invade and smother grass in derived savanna, forming dense thickets above head height. These also burn, but the stems remain standing and offer shelter to the soil. Hunters do not like the plant because the entangled stalks, even after burning, can make it difficult to pass through the area. When Eupatorium burns the fire is hot enough to kill young trees, which is a disadvantage for tree crop plantations.

Tree crop plantations are more vulnerable to fire than natural forest. Deciduous species, such as Gmelina and Teak, give a heavy leaf fall in the dry season, and the leaf litter can burn fiercely. Gmelina is less resistant to fire than Teak, and the soil heating may kill the cambium at the root collar. Another serious consequence is the killing of the undergrowth, allowing rain to beat directly on the soil and causing serious sheet erosion. This may remove soil to a depth of several centimetres, leaving roots exposed on the crest of a knife edge of soil. In order to improve access for tapping, peasant farmers commonly use fire to control weeds in their rubber plantations, because rubber is relatively fire resistant. Palms are also resistant to fire, when sufficiently grown, but repeated burning of ground vegetation can lead to soil deterioration and compaction. In exceptionally dry years, cocoa plantations and citrus orchards may burn through, killing the trees; and fire may invade the drier forests during drought years. However, in the high forest zone, it is beneficial if relicts of natural forest are retained in the matrix of forestry or agricultural tree crop plantations, along watercourses (say to 50 metres either side), and on swampy ground, and on steep slopes and the rocky tops of hills. This makes natural firebreaks, restricting the spread of fires and facilitating fire control. Moreover watercourses are protected from erosion or silting. They may also serve as a **genetic reserve** of natural species, composing ten to twenty per cent of the total area, depending on the character of the terrain.

Savanna plantations of *Eucalyptus* species are also vulnerable to fire, though they may resprout from the base of 'dead' trunks. *Eucalyptus citriodora* and the tropical pines are relatively resistant to fire unless the crown is burnt. If there is thorough clearing and mechanical cultivation prior to tree planting, the trees may close canopy and prevent the entry of grass and other weeds, so that fires are not a problem subsequently. However, grass which was well established before the trees closed canopy may not be eliminated. In fact mechanical cultivation is essential for successful establishment and growth of tree crops in the savanna zone, in order to promote infiltration of rainfall by loosening the soil, and to remove competing vegetation. In Sudan savanna, Neem (*Azadirachta indica*) is resistant to fire, and because of its dense evergreen crown, fire is not usually a problem in well established plantations of this species.

Conclusions

In the savanna zones, the demand for firewood, particularly by the towns, is denuding the landscape of trees. This has been made worse by the advent of motor transport, and the growth of urban populations. Average consumption of firewood is estimated to be more than half a tonne per head per year. Firewood may be carried from distances of up to 150 km from the cities, and charcoal perhaps twice as far. Removals are of the order of 60 tonnes of wood per hectare, leaving scrubby coppice regrowth with occasional remaining trees of **ironwood** species that are too hard to fell with native axes. The wholesale clearance of bush may extend from the roads to the distant horizons. In rural areas a farmer's wife may have to headload firewood several kilometres to her home. She does not believe in wasting money buying firewood, and if finding a supply becomes difficult, she will cut the shrubby regrowth, and after that dig out the stumps and roots. Finally she is reduced to burning dung and cornstalks, that ought to go to fertilise the fields. From being almost unheard of, this is becoming an increasing practice in drier parts of West Africa. Meanwhile impoverished townsfolk steal trees planted on public land, and may purloin them from other people's compounds if they get an opportunity.

The system of shifting cultivation has served Africa well for several thousands of years.

However it is now breaking down because of:

1 increased populations and growth of urban communities;

2 expanding cashcropping, as peasants try to progress from mere subsistence farming;

3 the tendency to try to over-exploit the land by removing farm trees, extending the period of cultivation by the use of artificial fertilisers, and by disregarding practices of customary land usage;

4 consequent reduction in the area of land available for free-ranging livestock and for forestry;

5 insufficient earnings to meet the demand for higher living standards, or to generate the capital needed for investment in agriculture;

6 inability to satisfy the needs of the towns for staple foodstuffs, meat and forest produce;

7 limitations in the pace of change, and the degree of change, that peasant farming can absorb without disrupting the fabric of society.

Traditional peasant farming conserved the soil, but did not improve it. Soil management was regulated by customary usage, and not by the thoughtful application of scientific understanding. Almost all that a farmer invested in the land was his own labour. Having now more or less joined the consumer society, people are no longer satisfied with a way of life that quite definitely confines them to the lower strata of society, and to relative impoverishment. Peasant farming is not, and cannot be, sufficiently remunerative to keep people on the land.

New farming systems are needed, that will boost the productivity of the land – for staple foodstuffs, for meat and animal products, for fuelwood and timber.

Such systems must:

1 be profitable;

2 keep the land in permanent, rather than intermittent production, by utilising the fallow or dispensing with it;

3 employ less labour or produce more food;

4 protect the soil from surface erosion, and from loss of the fine fraction of the topsoil;

5 maintain, or better still augment and deepen, the humus content of the soil;

6 augment and balance the supply of soil nutrients;

7 maintain soil tilth, and protect the soil from compaction or waterlogging;

8 regulate the use of water, to optimise consumption, and to extend growing of crops throughout the year;

9 reduce the risks of either soil leaching or salination;

10 encourage the activities of natural soil cultivators – the earthworms, insects and micro-organisms that live in the soil;

11 control the growth of weeds, and the incidence of pests and diseases.

These developments will require new men, who are willing and able to put new methods into effect. Although not necessarily large-scale, they will be on a larger scale than peasant farming. Moreover they must survive in the same environment as that in which peasant farming is beginning to fail. The new farming methods will be directed towards cheap production of bulk materials in uniform grade, suitable for industrial processing, and making possible high living standards for ordinary people.

The challenge is more difficult than that faced by European countries at the same stage of development, because of the greater fragility of tropical soils, and perhaps because African countries today are more exposed to world trade. The aim is to guarantee the continuity of increasing food supplies to enlarging urban and industrial communities. Farming has to be made less vulnerable to the ravages of nature, and to the misfortunes of time and circumstance; and farming systems are more likely to be robust if they are near biological balance. But, if man is to continue to *'Be fruitful and multiply, and to replenish the earth and subdue it: and have dominion over the fish of the sea, and over the fowl of the air, and over every living thing that moveth upon the earth . . . every herb bearing seed*

. . . *and every tree* . . .'[11] then the other changes that go with progress in agriculture must also be confronted.

11 *The Holy Bible*, *Genesis* chapter 1, verses 28 and 29

4 The land tenure impediment

> *'Land belongs to a large family, some of whose members are dead, some are living, and innumerable others are yet to be born.'*
>
> *A Yoruba Chief quoted by F.D. Lugard*

Man dominates the animal and plant kingdoms, despite his comparative lack of natural organs of offence or defence. His success is only partly due to his super intelligence and manipulative hands, but also because he is a sociable creature. The proverb – *United we stand, divided we fall* – is not just idle words. Personal advantage has to defer to communal welfare – though to be more realistic there is an uneasy compromise between the two. The important consideration for an individual is that *his* family and *his* children should survive and prosper; but they cannot achieve this without the help of you and yours, though at the same time you may all be competing for the same resources. To that extent you are a threat to each other's existence. So human communities make **laws** and **rules** to govern their own conduct; draw up individual **rights** and **privileges** which are not to be infringed; and appoint individuals or institutions to be their guardians. Laws and rights are less mutable; conforming to them is called *right*, and breaking them is called *wrong*.

These laws and rights may regulate either human behaviour or the allocation of resources. The aim is to reduce conflict within society, and to obtain harmony and cooperation. Land tenure, in particular, concerns the allocation of natural resources among the individuals of a community, both severally and communally. However the needs of communities change as they enlarge and expand – or as they contract and diminish – and also as their impact on the environment alters it in various ways. Moreover there is the effect of their intensity of contact with neighbouring communities, and the consequences that this may have for the way they organise themselves.

Some of the more basic tenets regulating human behaviour scarcely alter over long periods of time; witness religions and religious injunctions which are still being practised and have stood, perhaps, for millenia. They may reach back to a time even before humans began to practise agriculture, between 5000 and 10 000 years ago.

Human populations can modify their environment quite quickly, even within a decade, and sometimes profoundly; therefore those laws and rules that regulate the allocation of resources must be subject to continuous modification to suit new circumstances. Though people may not sharply distinguish them, a much wider interpretation is possible for the injunctions *'You must not steal'* or *'You must not covet your neighbour's belongings'*, than for *'You must not kill'* or *'You must not commit adultery'*. Your neighbour's belongings are open to redefinition, as is clearly demonstrated, for example, whenever Government imposes *'Pay as you earn'* tax, or increases it without the individual's consent. Land tenure is the allocation by a community of individual rights in a resource, and this *does* change with time, and *must* do so, if a community is not either to commit mutual suicide, or at least to cut itself off from progress and prosperity.

Influences of colonial and pre-colonial eras

The continent of Africa is still relatively lightly populated, with a general population density approximately equivalent to that of South America and Russia, about half that of the USA, and a seventh of that of Asia and Western Europe. Nevertheless, at the time of the 'Scramble for Africa' by the former colonial powers a century ago, this continent had only about 10% of the population it has today, whereas Western Europe had a little over half its present population. At that time, the main purpose of British imperialism was to safeguard and extend overseas trade to provide a livelihood for an increasingly industrial population. France and Germany, on the other hand, still had substantial peasant populations and could think in terms of settling them in Africa, though in fact this did not happen to any considerable extent. However a consequence of these different policies was that Britain secured the more populated parts of Africa, because trade was with people, whereas France in particular was content to spread her imperial rule over as large an extent as possible, including considerable tracts of desert. Britain's other interest was to protect her lines of communication with India and her Far Eastern possessions, via the Suez canal; and therefore to exclude potentially hostile rivals from strategic localities en route, mainly along the East coast of Africa.

A consequence of these various motives is that the anglophone countries of tropical Africa, by reason of their greater populations, and perhaps for reasons of greater political and economic sophistication, are the first seriously to feel the need to metamorphose from traditional to modern national economies, and to begin to suffer the painful socio-economic consequences of this impending trauma. These have become particularly evident since the countries received their national independence, around 1960. Indeed one wonders if this is why they suddenly became less attractive as colonies. The situation has recently been exacerbated by the world economic recession, and also by chronic drought affecting particularly the more arid parts of the region. Perhaps, in various ways, their governments have made the situation worse, both through a lack of understanding of the nature of the change, and possibly through a nostalgia that causes them to try to hold on to the past.

Prior to colonisation, much of tropical Africa was in turmoil. One may wonder if, in the early part of the nineteenth century, there had been some climatic change, such as we have seen since 1973, which forced communities into motion. Was Dan Fodio driven by drought as well as by religious conviction into northern Nigeria, Dahomey and Cameroon? Whatever the reasons, this pressure from the north set the various tribes into collision; and by the middle of the nineteenth century the consequences of it had reached to the coast of West Africa.

At the same time there was the disturbing effect of the opening up of the African hinterland to world influences. Heralding the change, religion – both Christian and Moslem – was also on the march, and trade quickly followed after. A consequence of these troubled times was that communities had to be more cohesive, if they were to defend themselves, and this was also reflected in the way they practised agriculture. In the savannas, the population lived in walled cities or stockades, emerging to cultivate their surrounding farmlands during the rains, which was usually observed as a sort of 'closed season' for warfare, because even soldiers have to feed. Where savanna peoples had encroached into the forest, they too lived in walled cities. Less civilised tribes occupied the tops of rocky hills, which provided natural fortifications – and they might add to these by hedging their compounds and fields with prickly Euphorbias, and huddling their huts and granaries together with interconnecting walls. Across the intervening expanses wandered the nomadic cattle peoples, grazing their herds and discouraging molestation by fierce courage, but providing a welcome source of milk, meat, leather products and dung to manure the fields. The forest peoples, on the other hand, lived dispersed, because the forest was their natural protection. The forest was also an obstacle to communication, particularly to travel by horse, and perhaps as a result they were less organised than most savanna peoples.

Conquerors took wives from among the conquered, and attempted to impose on the survivors their own laws, language and religion, but in turn absorbed to some extent the attitudes and customs of the defeated people. Polygamy, and the existence of domestic slavery, encouraged internecine warfare, in order to capture wives and slaves; Arab and European slavetraders exploited these circumstances, and made a bad situation worse.

The unsettled times caused communal farming to be much more usual than it has been since, both in the savanna and forest zones. A community, or descent group, farmed the same fields and stored their crops in communal granaries or barns. The men of the group worked their fields jointly under a leader. This promoted the group solidarity that would be needed in battle, and also protected the community from the dislocating effect on farming operations due to individuals lost in war. Again, as a precaution against crops being despoiled or looted, farming might be in several widely separated fields. These would be intercropped in the usual way, and allowed to recover under a bush fallow. However, after 'pacification' by the colonial powers, the need for security was diminished, and the populations began to disperse. Farming hamlets that had been occupied only during the farming season, became permanently inhabited villages, and some grew into small towns. In some walled cities, the populations diminished and the walls were allowed to fall down or were used as a source of building materials – although the numbers of citizens living in the town fluctuated widely, increasing in the dry season and diminishing in the rains when the people scattered to their farms. At Ibadan, the population in the dry season could be twice that in the rains.

The former pagan tribes descended from their rocky eyries, sometimes forced down by the colonial administration who wished to curb their warlike propensities, and they began to farm the more fertile and more plentiful soils on the unprotected plains. The nomadic herdsmen were able to roam more widely than before, sometimes into new areas such as the Cameroon highlands. Everywhere, the more enterprising individuals were able to escape the confines of tribal society and to seek their fortunes in commerce and industry far afield. Before the days of national governments in Africa, and the advent of passports and immigration formalities, people began to travel widely and regularly throughout the region in which they were born, for example from one end of the West African Coast to the other, and from Lake Chad to the sea, and similarly in East Africa.

Development of individual proprietorship of land

Quite quickly, where group farming existed, it was abandoned in favour of individual cultivation. Cole, writing of the Kagoro district of northern Nigeria, describes the **kindred** farming of pre-British days, and states, 'Nowadays the merest family tiff will result in the individual demanding his portion of the family lands and going off to farm on his own.'[1] The common fields might persist, but each farmer was allocated his own piece or pieces to cultivate, which might be located in the various fields; he stored the fruits of his labours in his own personal granary or barn. In this way he, his wife and children could gain the maximum benefit from their own efforts. A farmer might be granted a different patch of land each time a field was returned into cultivation after fallowing. Clearly, because the careful farmer would leave his plots in better condition than a negligent farmer, he would prefer to come back each cycle to the same patches of land. These would subsequently pass to his heirs. Both systems still survive in some of the more southern parts of Nigeria, such as Owerri and Ikot-Ekpene in the East; and Lloyd also records communal farming at that time at Ijebu-Ode in the West.[2]

The more densely populated parts of the country, sometimes with over 400 persons per square kilometre, have been the most conservative in their systems of land tenure and arable cultivation. Indeed it could be said that the common field has been retained as a sort of 'dole' for what are effectively the landless members of the community, to be divided up and allocated amongst those eligible at each farming cycle.

Africans, in general, have a strong sense of personal possession. It is widely held that a man and his heirs may benefit from his own improvement, whether it be clearing virgin bush, planting fruit trees or building a house, provided that it is not in the family compound. Clearing new land is a considerable labour, particularly in high forest, and it is more exposed to depredations by wild animals and vermin.

1 Cole, C.W. (1952) *Report on land tenure Zaria Province*, Government Printer, Kaduna (p. 84)
2 Lloyd, P.C. (1962) *Yoruba land law*, Oxford University Press (p. 378)

He would, of course, require permission from the community or family head before initially clearing the land, but this would not normally be refused and no payment would be expected, unless possibly a token gift. The land would continue to be his so long as he continued to use it, and did not either abandon it or allow it to remain unused for much longer than a normal fallow – say more than ten years – or allow someone else to farm it without protest. He could lend the land or rent it to someone else to farm, and he might do this purposely if he were unable to farm it himself, perhaps because it was too far from his other fields or because he had too much other work to do, in order that his proprietorship should continue to be acknowledged. He could not, however, sell the land, because ownership remains vested in the community, and he could not alienate his rights without permission from the head of the land-owning group. With permission, he could exchange the land, perhaps so as to consolidate his holdings; he could also pledge it, borrowing money in return for the lender obtaining rights of usufruct from the land until the debt was repaid (in effect as a sort of interest on the loan). The lender might either cultivate the land himself, or put it out to rent (which was more usual if he was a trader). The period of the loan would be unspecified, but would have to be at least long enough for the lender to be able to harvest any crops he had planted; but whenever the debt was repaid, and after however long, the land reverted to the original proprietor.

Because of the need for fallows, a man may mark the boundaries of his land. In the forest zone he can use a distinctive shrub or tree, such as *Newbouldia laevis* or *Dracaena* species, both of which can be struck from cuttings; in the savanna he may rely on a ditch or bank. A man may hold land both in the communal field and in individual holdings, the latter being either inherited or cleared by himself. Even fifty years ago, except in some more densely populated areas, unallocated virgin bush was available, and often plentiful, and a man could obtain as much land as he could conveniently cultivate. In fact, most land being farmed under traditional tenure in West Africa is as individual holdings on family or community lands. People are reluctant to move on to land without permission from those whom they regard as the traditional land owners. However, in parts of northern Nigeria, there were extensive unpopulated or depopulated lands, for which traditional land owners no longer existed. In the early part of the twentieth century people, such as those freed from slavery, took up occupation of these lands without challenge, and established themselves and their descendants on them. Indeed, this is permissible under Muslim law.

The most significant change during the twentieth century has been the shift from subsistence farming to cashcropping; that is from farming mainly to feed the farmer and his family to farming deliberately to earn money. Cash crops were formerly export crops, such as cotton, groundnuts, cocoa, cola and rubber, depending on local climate and soils. However, today, food crops may pay better and export crops may give way to cereals, cassava, yams or citrus. The farmer frequently hires labour to help increase the size of his farm. The labourers are often young men, who would rather earn money than help their fathers for nothing; or they may be strangers (sometimes other nationals) coming from drier more northerly districts, where the rains have not yet begun. Perennial crops have effectively changed land from communal to individual ownership. Cocoa continues to yield for several decades and, though in theory the land ought then to return to the common pool, this is unlikely to happen. If a man sells his improvement, for example, his cocoa trees, to another man, he is in effect alienating the land. He may otherwise pledge the crop, then the lender can continue harvesting the pods until the debt is repaid. However a consequence of cashcropping, whether of tree crops or arable crops, has been a rapid diminution in the area of unallocated land, so that nearly all the land has passed to individual proprietorship.

All land is regarded as belonging to one community or another. The head of the descent group, or the chief, is the trustee for the whole group. He allocates land from the common pool to individuals and their heirs, so long as they continue to cultivate it. If there is a failure of heirs, or it is left uncultivated for a certain period, it reverts to the common pool. However people are reluctant to terminate a man's right to a piece of land so long as he continues to assert his proprietorship and does not allow others to usurp his occupancy. If the proprietor wishes to allow others to use his land, he will be expected to notify the family head or chief, and he will usually be anxious to do so in order to make it quite clear that he wishes to retain his proprietorship.

Inheritance of rights in land

As has already been explained, rights in land may be inherited. More usually only men exercise proprietorship over land, and women have rights to use land only through their husbands or their fathers. It is possible that matrilineal inheritance, as amongst the Ashanti of Ghana, was formerly more widespread, although today patrilineal inheritance is almost universal. An explanation offered has been that, due to the continual loss of men in war, the women became effectively the occupiers of land. In matrilineal inheritance, a man's brothers inherit, not his own children. Failing living brothers, his rights in land would pass to the brothers' children, or failing that to the mother's uncles or their offspring. However, in Ghana, the cultivation of cocoa has caused a modification to these traditional laws of inheritance. The subject of bequeathing cocoa farms was first brought up to the Council of Chiefs of Ashanti in 1938, and reopened again by the Asantahene (the suzerain) in the 1941 session. 'One fact with us Ashantis is that we appear to be too conservative. We always like to stick to custom even though it may have outlived its day. I do not deprecate the idea of brothers and nephews succeeding their deceased brothers or uncles, neither do I propose that the custom should be abolished, but I want you to understand that our children are blood of our blood and bone of our bone for whom we are accountable to God for bringing them into this world. God and our country expect us to make provision for the children we bring into this world. They and their mothers help us in our farm and domestic work. Sometimes you find that all your nephews and nieces do not come near you at all. It is only your children who care for you. Is it not fair then that we should make provision for them and their mothers who look after our interests and welfare, so that they may not become useless and wretched after our death? I would advise you to consider this question very carefully and favourably.' This matter came up again for discussion in 1942, when the Asantehene proposed a variation in the existing custom, so that, 'When a person makes a gift of his own personal or self-acquired property to his children or to any other person in the presence of accredited witnesses, whether the relatives of the donor approve of it or not, it becomes valid.' After lengthy discussion this was accepted by the chiefs.[3]

In patrilineal inheritance, the land is inherited by a man's own sons, and if he has one wife it is divided equally between them. If the deceased is polygamous, the land is first divided equally between the several groups of full brothers, and then equally within each group of brothers. That is the child of a wife with many sons will receive a smaller portion than the child of a wife with few sons. No land would be allocated in respect of a wife who was childless, or who had only daughters. Amongst some communities, as in some parts of both Hausa and Yoruba, daughters inherit equally with sons, but this is not

3 Busia, K.A. (1951) *The position of the chief in the modern political system of Ashanti*, Oxford University Press (p. 233)

usual. Widows do not inherit land, but are themselves 'inherited' by a younger brother or a son of the deceased, who will be responsible for them, and through whom they will get continuing rights in land; sometimes they may go to men of their choice within their husband's family, if there is not otherwise a suitable match. For a woman beyond childbearing age, or who passed into the care of her own son, it would of course not be a consummated marriage. If the children who should inherit the rights in land were minors, the land would pass into the care of an older brother or uncle. He could use the land for as long as it was in his trust, but he would be expected to relinquish it as soon as the heirs were old enough to require it for themselves. An older brother cannot normally inherit from a younger brother; in the event of the death of the father, or of a younger brother, the older brother will look after his younger brothers or their children properly and without the temptations of covetousness.

If a man, before his death, indicated a preference to bequeath his land to a particular child, or to someone other than his own offspring, this would be disregarded and the land would be subdivided in the traditional manner. Of course, during his life, he could give land to his sons as he wished and this might not be challenged after his death, provided that they had already taken up occupancy of the land. Indeed his sons would expect to be allocated a portion of land on marriage, and a young unmarried man might be given a plot of land to farm. In some communities, the eldest son inherits a greater share of the land, because of his greater responsibilities to the family.

Where women inherit land equally with men, a man may prefer to gravitate towards his mother's family rather than to his father's. The reason is that women will tend to have more land at their disposal than men, because unlike a man the number of offspring of a woman is limited to the number to which she can personally give birth, and a child will only share his mother's land with his **full** brothers and sisters; on the side of his father he will also have to share with his **half** brothers and **half** sisters, if his father was polygamous. Where women do not inherit equally with men, a woman's property may pass to her daughters and a man's property to his sons.

Acquisition of rights in land by strangers

With the acquiescence of the community, the family head may allocate land to strangers. The strangers may be from different communities of the same tribe or, less commonly, from other tribes. The stranger in effect joins the community and must recognise its head. His rights in land are exactly the same as for other members of the community, and his heirs inherit his rights in just the same way as do those of the other members of the community. He is not allowed to alter the use for which he acquired the land – either to cultivate or to build a house. He would not normally be provided with land to plant cocoa, which would effectively alienate the land from use by the community, though ultimately, when he is accepted as a full member of the community, he might plant a tree crop without challenge. Small communities are pleased to add to their size, and thereby importance, provided that the new members are acceptable and prove to be social assets.

All the members of a community formerly paid a tribute to the chief or family head, or they might work on his fields for him without pay. Today these practices are often discontinued, because the farmers pay tax, and the chief may receive a government stipend which is regarded as being in lieu of the former traditional gifts. However, when he allocates land, or assists in settling a dispute about land (or other matters) he will receive a gift, though this will by no means represent the value of the land, but is a customary mark of appreciation.

The nomadic pastoralists have no rights to land. They may present the village head with gifts, and may pay cattle tax to the local government treasury. Their presence may be an asset, because they can provide milk, meat and skins in exchange for the products of cultivation; the dung from their cattle helps to manure the fields, and they may be welcome so long as their cattle do not damage the crops. Of course there is sometimes violent conflict between them and the peasant cultivators, but the latter may fear to molest them because they can react with ferocity, and being of no fixed abode they are not easy to bring to account without the cooperation of their chiefs. The extension of the cultivated area is reducing the bush available for them to graze their cattle, particularly during the growing season. During the dry season the formerly cultivated areas may be used as pasture, but will have a reduced capacity for fodder compared with natural bush. Certainly, if the nomads wished to settle, and land were available for them to cultivate, they could be allocated land just as would any other stranger. However it usually requires some catastrophe to induce them to change their style of life, such as the loss of most of their herds due to severe drought or to epidemic disease. It may also be noted that the value of a cow is approximately equal to that of a hectare of land, so that by selling half their herd they could acquire sufficient land to maintain the other half, at least with improved pasture.

Development of freehold tenure

For more than half a century, there has been a trend towards **freehold** land; this is land as a freely negotiable asset, which is owned by an individual, disencumbered of communal rights, and which may be distrained for debt. The concept has its origins in the towns, where it enables the investment of capital in building a 'modern' cement house. This has several purposes, because the owner will be a person who is saving money from his activities in trade, or industry, or contracting work. It provides him with somewhere to reside, with rooms to let as shops or living accommodation giving an additional source of income, and also furnishes collateral for borrowing money. Besides being a personal convenience, this adds to the owner's status. A building is an important asset, underpinning his business and making it possible to gear up his commercial activities when needed, by taking loans. These may be by mortgage, or by character lending based on his manifest ability to accumulate capital, thereby being creditworthy.

The concept is supposed to have been first introduced in Nigeria by returned slaves, at Abeokuta, which was then an important trading centre *en route* from the coast to the hinterland in the middle of the last century. Following the imposition of colonial rule, at the turn of the century, the purchase or alienation of land from communal to individual ownership for the purpose of building a house became usual in most of the major cities. This was doubtless promoted by the burgeoning class of lawyers, trained at the English bar, who introduced English concepts of ownership. Also there was a dual operation of English law, associated with colonial activities of administration, commerce and criminal justice, and of native law and custom which still applied in civil matters pertaining to land disputes, divorce etc, with regard to the local people. However the rapid spread of the concept makes it quite clear that it fulfilled an existing need; it could also have arisen quite spontaneously as it did in Europe, and for exactly the same reasons. Formerly, when houses were constructed entirely in local materials, a man would not have contemplated buying a house – he would have built one for himself.

In the Abeokuta district (Egba) the concept of freehold had also spread into rural areas, so that a man's farms or cocoa plantations could be distrained for debt. It must be remembered that Abeokuta was a city of refugees, founded in about 1830, and that, except for a minority of the people, the land could by no means be regarded as ancestral property. However, until recently, in the rest of Yoruba land and for much of the rest of the country it was inconceivable that farmland could be alienated from the community, or bought and sold. As a Nigerian chief explained it, 'Land belongs to a large family, some of whose members are dead, some are living, and innumerable others are yet to be born'. The land proprietors could commit no act that would deprive their descendants of the right to enjoy the land. However, since 1970 and consequent on the oil boom, a flood of money has been generated in government and in commerce and industry, and has activated a general desire to acquire land both as a business asset and, because of the increasing price of food, for the possibility of farming. The only way to free land of communal rights is to purchase it from a neighbouring descent group, which has the merit that the purchaser is more likely to obtain a valid title by knowing those who should be approached. The head of the descent group acts as a trustee for the disposal of the land, but he may not sell it or otherwise alienate it from the family without at least the tacit consent of the community as a whole. If he did so, at least in Nigeria, the sale would not have been regarded as valid. Nor may an individual farmer, even though he possesses exclusive rights to use a piece of land, sell it or alienate it from the community without the consent of the family head and the representatives of the community.

Plantation agriculture and land tenure

Lugard wrote, 'The cultivation of exportable products by natives on their own holdings is unquestionably preferable to a system of European owned plantations worked by paid labour, provided that the enterprise does not require large capital or technical skill and management, and that the crop is an indigenous or naturalised one. The native holder occupies a higher status, and working in his sole interests and in his own time, will produce more than if working for another, as we have seen in the palm produce, cocoa and groundnut industries. On the other hand, it must be admitted that, though the cost of European supervision is heavy, it is compensated for by improvement in the quality of the output, and by continuous returns due to better care of the soil; but European ownership on a large scale has in all cases resulted in the demand for alien or compulsory labour, by which alone large foreign-owned estates can be kept going'.[4] In fact, in the former British West Africa, the creation of European-owned plantations was not encouraged (in fact it was actively discouraged) by Government, and in the western part of Nigeria for

4 Lugard, F.D. (1922) *op. cit.*

example, this was reinforced by the fact that the territory had been ceded by treaty, and not taken by conquest, so that traditional tenure was intact. The operation of traditional tenure made it difficult for strangers to obtain exclusive title to extensive tracts of land. The few plantations that existed were mostly relicts from grants of land predating the First World War.

In former French and German territories, attitudes were different. Being less populated at the time, there were large areas of more or less vacant land, on which the colonial governments could settle their nationals. However, before the advent of modern medicines, West Africa was extremely unhealthy and often fatal for Europeans, and European-owned plantations were relatively few.

Individual rights in communal tenure

There are no means for an individual to buy out the rights of the other members of his community to a piece of land which he is farming; there is, for example, their reversionary right to the land should he fail to continue exercising his existing rights of cultivation, or for that matter, if he were not to build a house if the land was granted to him for that purpose. House building takes precedence over agriculture, and land adjacent to a community will be given out for this purpose, and the former cultivator compensated with land further afield – though possibly to his displeasure at the disturbance. If a man is allocated family land to build a house, and he fails to build on it within a reasonable time, it will be reallocated to somebody else, notwithstanding the traditional gifts that he may have donated to obtain the concession. Once he has built on the land, it continues to belong to himself and his heirs, even if the house falls down, so long as the foundations remain visible, or unless it has clearly been totally abandoned, whereupon it returns to the common pool. However, if a man *buys* land, it is his to do with when and as he likes, subject to the limitations of the laws of nuisance or of town and country planning.

A family may, by agreement, subdivide land between two or more branches of the family – that is, between brothers or their respective descendants. One branch will inherit the new family headship, and (normally) the eldest male members of the other branches will become family heads of their respective segments. This could happen owing to discord, or irreconcilable conflict within a family, or because the family had simply become too large and unwieldy. It would tend to occur on the death of a family head. Usually a family head is replaced only on his death, though clearly if he became conspicuously mentally incapacitated he would be replaced. Though preferment is usually automatic by reason of age, a worthless individual might not be permitted to become family head. Clearly the eldest individual is the person closest to the family ancestors, and is therefore the most suitable representative of both the living and the dead, and the one who will be soonest reincarnated in his own descendants. In principle there is no reason why family land should not be subdivided by agreement down to the last completed generation – i.e. that which cannot be added to by further births.

Where a man buys extensive tracts of land, on his decease these may effectively be treated as land communally owned by his own descent group. When wealthy men have died it has not been infrequent for there to be protracted lawsuits between his wife and children on the one part, and the family group to which he belonged on the other part, disputing the administration of his estate. As already explained, it is contrary to customary law for a woman to inherit or become trustee of her deceased husband's property. This may be deeply resented by a woman, particularly where she has made a significant contribution to their joint wealth. Clearly this is an area in which modern thought is modifying traditional practice; although new mores are not yet firmly established.

The system of inheritance of rights in land results in fragmentation and dispersal of holdings. A man may inherit parcels of land from his various ancestors, as they cleared new lands on the margins of existing farmlands, and he may likewise have added to these himself. The inherited lands will have been subdivided with his brothers, and previously with his father's brothers and his grandfather's brothers, and so on. In addition he may have rights in a communal field, if such still exists. Among people where women also inherit, he may have acquired access to parcels of land belonging to another community. These lands may also be subdivided among his own heirs, who may insist on subdivision of some of the choicer locations, such as those nearer the town or those bearing perennial crops, rather than accepting allocation of discrete blocks. The result is a complex patchwork of rights in land, which may be inimical to its efficient use. Indeed, some of the land may be rented out simply because the farmer is unable to operate over his scattered plots, although otherwise he might himself prefer to cultivate a larger area of ground.

The difficulties are compounded by the rights which other members of the community may exercise over land that belongs exclusively to an individual to cultivate. It is not unusual for indigenous fruit trees on the land to be reserved for the farmer's own use while he is actually cultivating it; and this is commonsense, because other people coming onto the land to harvest from the trees might damage his arable crops. However, when the land lies fallow, any member of the community may be permitted to visit the trees. In other instances, rights to usufruct of trees may belong to individuals other than those that cultivate the soil. This may occur particularly where the proprietor has lent or rented the land to another person to raise arable crops, but has retained the right to harvest the trees himself. During the dry season, the livestock of the community may be permitted to roam at will, and as the fields are unfenced the animals cannot be restrained within, or excluded from particular areas and the farmer could not claim damages from the owners of the animals – though during the growing season this would be another matter. In the savanna areas it is not improbable that nomadic herdsmen would break down fences erected across their normal lines of access. Hunting or fishing across the entire lands of the community may be permitted to any member who wishes to avail himself of the opportunity, though recognised hunters may inherit the vocation from their fathers, and belong to a special guild or society. Their activities are regulated by ritual and tradition, and formerly they also served as scouts in the defence of the community. Today the right to hunt may have been assumed by every person who can acquire a gun, whether native or stranger. Although strangers have no automatic right to hunt on the lands of a community to which they owe no allegiance, nevertheless it can be difficult to exclude them.

Development of tenure in Northern Nigeria

In the north of Nigeria, the traditional forms of land tenure of the Hausa closely resemble those of the Yoruba and of the southern parts of West Africa, as already described. The trends affecting land tenure in the south were also occurring in the north, perhaps because town dwelling was already a marked characteristic of the Hausa people. However, the evolution of land tenure in Hausaland has been disturbed by conquest: firstly by the Fulani invaders in the early part of the nineteenth century, and secondly by the imposition of British rule at the beginning of the twentieth century. The former resulted in traditional land tenure being overlain by Islamic law, and by Moslem rights of conquest. The Shari'a concept of absolute ownership of land conflicted with the customary law which guarded community rights. This is not entirely resolved. Unoccupied lands became, at least theoretically, vested in the Emir (the suzerain) and, via his local representatives, could be disposed of at his pleasure to whomsoever he wished. This prerogative was usurped by the colonial administration as a right of conquest.

According to Islamic law there are two kinds of conquered lands: cultivated land and waste (or 'bush'). There are two alternative rules for cultivated conquered lands: the ruler can keep them as the property of the State, and put them to use taking the proceeds into the State treasury; or he can divide them among his warriors, each of whom may sell his own portion, or give it to someone else, or keep it for himself and his heirs to inherit. A man may give family land, or rent it, or keep it for his heirs and the ruler has no say in the right of ownership. Territory ceded by treaty (i.e. without fighting) belongs to the original owners, and not to those acquiring it. Bush near the town cannot be taken without the prior approval of the ruler, because the land is needed as a source of firewood and of grazing for animals. Whoever first takes occupation of premises in a conquered country possesses them, but he cannot sell or rent the property so long as the buildings put up by the enemy remain erect.[5]

Besides the Emir and his family, land was given to **fiefholders**, and these lands were worked by slaves and labourers. The fiefholders gave specified proportions of the harvest to the Emir. When a fiefholder died the land reverted to the Emir, who reallocated it. All farmers paid a **tithe** (one tenth of the produce of the land) to the Emir. With the advent of British rule, the fiefholders, the status of slave and the office farms (i.e. granted to various officials under the Emir) were abolished. The various exactions on farmers were replaced by a head tax levied on adults of both sexes. The colonial administration also usurped the general rights over lands formerly belonging to the Emirs, and over the personal estates of deposed chiefs. The slaves were absorbed into the community, and became as other peasant proprietors. However the Moslem community, unlike the former pagan communities, was not obliged to furnish lands to all its members, and in densely populated areas it was possible for individuals, by force of circumstances, to become landless.

The British were anxious to prevent abuse of land ownership, with which they had been familiar in India, and perhaps to some extent in their own country. They wanted to avoid creating a class of landlords, of share cropping, or rack renting, of land speculation, and of the growth of landless populations in circumstances where making land available for all citizens was the only kind of social security. In the north of Nigeria, the Fulani chiefs, who 'owned' towns and large tracts of country, were replaced by salaried 'district heads', and this

5 The source is *Muhammadu, Alkalin Zazzau*, quoted by Cole, C.W. (1952) *op. cit.*

eliminated the only class that could possibly have been called landlords.[6] The attitudes of the colonial government were essentially conservative; indeed this was an inevitable attribute of 'indirect rule'. Their natural sympathy was with the traditional and rural, rather than with the townee and radical. They were anxious to avoid alienating people from their traditional background, and for this reason tended to view the activities of the Christian missionary with an uneasy eye, and perhaps to favour Islam where it was already established. The British presence was small, and the administration economical if not parsimonious; colonial government placed a high priority on having a quiet life, and avoiding anything that might generate strife and discord – even if this ran counter to the needs of development. They were not opposed to development *per se* – very much the contrary – but the quickest way for a colonial servant to put himself in disgrace would be either to cause a riot, or to fail to quell it efficiently without spectacular loss of life! They could countenance a degree of oppression and injustice, an inevitable corollary of indirect rule, provided that it was not too obtrusive or painful to their consciences. The intentions of the *Land and Native Rights Proclamation No 9* of 1910 were therefore startling, and, according to Lugard, nationalised the land. It was based on a theory of 'economic rent': that the land is the ultimate source of all wealth, and its ultimate ownership should be vested in the State, in order to secure, for all, the 'economic rent' and the 'unearned increment' arising from the increased value caused by development, whether due to expenditure by the State or by the community. Land is nationalised when the economic rent is paid to the State and not to an individual. Lugard was not too sympathetic with these revolutionary ideas, and wrote, 'That so great a change could be effected without difficulty, in spite of the growing recognition of private rights in land to which many witnesses bore testimony, may be ascribed to two principal causes. In the first place the local population were entirely ignorant of the fact, since there were probably not half a dozen Natives of Northern Nigeria who could read and write in English, and in practice no attempt of any kind was made to issue titles, or to interfere with the existing regime. In the second place, much of the land in Northern Nigeria, except in the vicinity of the great cities, was unpopulated jungle . . . and ample land was always available . . . Moreover, the peasantry had for generations been accustomed to despotic rule, while the rulers had for the most

6 Lugard, F.D. (1919) *Political memoranda*, *Memo No X – Lands*
The 1910 Proclamation owes its existence to Sir Percy Girouard, a railway expert and Governor of Northern Nigeria 1907–1909.

part lately been conquered, and were glad to accept the conditions imposed by Government . . .'

The law enacted that no title to occupation should be valid without the Governor's assent, and he was precluded from granting a right of occupancy except on rental. In effect all occupiers of land became State tenants. As it was at that time bureaucratically impossible to implement this by issuing certificates, it was a dead letter so far as the native occupiers were concerned. Subsequent alteration of the law and changes in practice maintained the right of native occupiers to the lands they occupied. The Government had to take regard of Shari's law and customary law as it existed in the local areas, so that the occupiers retained their former rights to the use and control of land against all persons except the Government. The Government held the power to grant titles to the occupation and use of land. It could eject occupiers who failed to pay their tax, or attempted to alienate their rights in land without the State's knowledge and consent. Lugard wrote, 'To exact a rental is to create the new and strange idea of property in the land itself' – Government would become the owner and not the trustee. In fact a rent on land was never imposed on the customary land occupier, but a head tax was – in effect a tribute paid to Government.[7]

The situation was reconfirmed by the Northern Nigerian Land Tenure Law of 1962. Rights to occupation and use of land were classified as either statutory or customary rights. Statutory tenure was leasing by Government for a specified number of years, on issue of a Certificate of Occupancy to an individual or corporate body. This could not be transferred or alienated by the lessee without the permission of the Government; the certificates could be revoked if the occupier failed to abide by the conditions of his tenancy. Government could compulsorily acquire, or dispose of, any vacant lands, and also of occupied lands provided that compensation was paid to the original occupier. On the other hand, customary rights of occupancy were vested in an individual member or members of a community according to native law and custom, and in effect were held by that individual and his heirs for an indefinite period. Depending on the locality, the individual could lend or pledge his land, or give it for rent on an annual basis, but he could not sell it. If land was transferred, the right of

7 Lugard, F.D. (1922) *The Dual Mandate in British tropical Africa*, Blackwood, reissued Cass, 1965 (p. 643).Lugard uses the term 'Native' to mean a person born in a particular locality, of parentage of the same locality; Africans from elsewhere he terms 'Aliens' or 'non-natives'. The words are used in their literal sense.

the original occupier to be compensated for 'improvements' was recognised; but as explained below, payment for improvements is not in essence different from payment for the land itself. In one respect the Government was endorsing a universal concept of traditional African customary law – that of inalienability of interests in land.[8]

Abdumalik writes, 'A result of "The Indirect Rule" through the Emirs and chiefs which the British used in the North [was that] the indirect rule made some emirs and chiefs so powerful that they discarded with following Islamic Law and proceeded to cheat their subjects. It went to the extent that a person was not even aware of his right of ownership on his house let alone his farm. In some places, immediately a person who owned a house died the house would be allocated to another person and that other person would eject the deceased's children and wives even before the completion of their iddah-waiting period. So, also the deceased's farm . . .' He continues, 'A Northerner is used to waking up in the morning and seeing surveyors coming to construct roads in his compound and to be shown the edge of his house. And if the construction affects his whole house, he will only be told "your house has been acquired". He has nothing to say other than that he and his family end up squatting in a brother's house until he gets another place . . . Sometimes, it was during a usual [i.e. ordinary] conversation that a person would be informed that a new road to be constructed would pass through his farm. It was not a matter of concern to inform an owner of a farm that a road would pass through his farm let alone to talk of paying him compensation. These irregular activities continued during that period.'[9]

Recent trends in land tenure

The Nigerian Land Use Decree of 1978 appears to attempt to impose the regulations governing land tenure in the northern States of Nigeria over the remainder of the country. It gives to a State Governor the right to issue or revoke certificates of occupancy, and to charge rent on such lands. Lands could not be subdivided, nor rights in land transferred, without his consent. In this situation, it could be

8 Famoriyo, S. (1979) *Land tenure and agricultural development in Nigeria*, Nigerian Institute of Social & Economic Research, Ibadan (p. 146)

9 Abdumalik Bappa Mahmud (1982) *Land in accordance with Islamic law and decree*, Report of a National Workshop, *The Land Use Act*, University of Lagos, May 1981 (pp. 23–24)

impossible to distrain on land, and real estate becomes a dubious security for credit. The provisions of this Act were enshrined in the national constitution of 1979, and cannot be challenged in a court of law. A consequence is that the law cannot be modified in the light of experience. Its provisions were hotly resented in the southern states of Nigeria, and every means taken, both legal and otherwise, to evade its effects. For example transfers of land are backdated to before the coming into force of the Decree in 1978, so that it is not applicable to them. It does appear, in many ways, to be an attempt to freeze the evolution of land tenure. Government is right to be afraid of the consequences of allowing the growth of a landless and rootless proletariat, and is understandably nervous of taking on the task of contriving new ways to cater for their welfare. In some countries there may be a tendency for militant Africanism to try to conserve what is seen as essentially African, rather than perhaps to see it as a stage in the continuum of development, which has been undergoing progressive modification as altering circumstances demanded change.

Writing in 1922, Lugard observed, 'The Government should not intervene between the native rulers and the people under their jurisdication as to the disposal of lands or the tenure upon which they are held, except to advise, and to maintain justice. Land rents should not be charged to natives, since Government is the trustee for and not the owner of the land and rentals are opposed to native law . . . The Government may also develop, for the public good, lands which are not in beneficial occupation, and not likely to be required for such purpose.' Herein, of course, lies the conflict in his views, where all land whether occupied or not, is claimed by the native community as belonging to them. Lugard continued, 'A system of land rentals, leases and licenses to native occupiers, and of the expropriation of holders without compensation except for their own improvements, appears to me not only incompatible with the principle of ruling through the chiefs, and the system of taxation recognised by the people, but also to violate the cardinal tenets of native customary law. It seems preferable that the natural evolution of land tenure should not be arbitrarily interfered with, either on the one hand by introducing foreign principles and theories not understood by the people, or, on the other hand by stereotyping by legislation primitive systems which are in a transitional state. Each advance should be duly sanctioned by native law and custom, and prompted by the necessities of changing circumstances.'[10]

10 Lugard, F.D. (1922) *op. cit.*

In Francophone Africa, matters have been somewhat different. The French have always been much more inclined to accept the supremacy of the State than the English, to whom devolved and parochial rule is an ancient prerogative. Moreover the lands they annexed were much less populated, and therefore there was less resistance to the alienation of land for European plantations. This development is particularly seen in the Ivory Coast and in the former German colony of Cameroon. The French were also much more inclined to impose their own values, which included the concept of freehold land. However they were also compelled to the expedient of indirect rule – that is governing through the indigenous rulers - in order to bring within feasible limits the expense of governing a vast imperial domain. But, as Lugard wrote, 'Upon the general acknowledgement of their right to allocate land, and to enforce punishment in respect of it, depends the prestige of the chiefs. Hence the system of ruling through the chiefs depends on the recognition by Government of these powers. A chief acts as a trustee for the tribe in regard to land.' He quotes a deputation (from Ghana) on the Gold Coast Forest Bill as saying that a chief 'is joint owner with his people, and he cannot exercise any proprietory rights without the cooperation of his people.' He continues, 'Consequent on these assumptions, it is maintained that every acre of land is the property of some tribe, family or individual, including forest and swamp.'

Francophone countries have not been immune to these constraints, and African customary law is widely applicable, and certainly regarded in the same light by the rural African as it is in Anglophone territories. Francophone governments are able to assume a much more authoritarian role than Anglophone governments may feel able to impose, or succeed in imposing; but even Francophone Africans have limits to their patience. In Francophone Africa, lands that have never been brought into cultivation are regarded as being at the disposal of the State, which was certainly not the case in southern Nigeria.

Forestry and land tenure

Professional foresters first came to British West Africa with experience from India and Burma. They were already aware of the need for forest reservation and control of exploitation, and knew how to obtain the silvicultural knowledge necessary for managing and regenerating tropical forest. The first Forest Reserve was created in Nigeria in 1899, the first Forestry Ordinance was promulgated in 1901, and the foundations of the Forestry Department were already established before the beginning of the First World War in 1914. (Contrast this with the creation of the Forestry Commission in Britain in 1919.) Indian experience had demonstrated that a well-populated country needed to dedicate about 30 per cent of its total land area to forestry. In Nigeria a target of 25 per cent was originally proposed as the proportion of land to be reserved for forestry.[11] The long gestation period needed for timber crops, and the low return on capital, discourage private investment in forestry. Many countries have met this need by creating a national forest estate. In most tropical countries it has been by legislation, setting aside delineated areas of land exclusively for forestry purposes. (In Britain, Government has purchased land on the open market for these purposes.) This enables the direct control of exploitation and regeneration by Government, and prevents alienation of the land to other forms of land use.

In Nigeria, as in most Commonwealth countries in the tropics, the constitution of forest reserves involved elaborate procedures for consultation between Government and all interested parties, including the different arms of Government itself, as well as local communities. The boundaries were actually marked out on the ground as the official inquiry proceeded; their existence was confirmed by a senior officer, who travelled round them. Local communities were visited in turn, and the purposes of reservation explained to them. Persons affected were encouraged to voice objections or to request continuing rights and privileges within the reserve, such as hunting or fishing, gathering wild fruits or snails for consumption, obtaining materials for rope making or other domestic use, and rights of way by named tracks – in other words to continue their former practices provided these were not inimical to the purposes of reservation. Compromises were reached; but rights differed from privileges in that the former were inalienable whereas the latter were discretionary, and could be revoked later if they conflicted with the management objectives of the reserve.

These proceedings took at least 18 months, and commonly several years. During the interim the proposed reserve was gazetted as a **Protected Forest**, to prevent encroachment before the *Final Order* was published as a Government Notice, which gave the full legal status of **Forest Reserve**. The gazette notice comprised two schedules: the first schedule gave an exact description of the boundaries of the reserve, in terms of bearings and distances, though

11 Lugard, F.D. (1919) *op. cit.*

rivers and streams, and (less satisfactorily) roads and footpaths might be utilised as boundaries; the second schedule listed the continuing rights permitted within the reserve, naming the categories of persons or communities to whom they were allowed. Farmlands were excised from the reserve, or the community agreed to abandon them, or was compensated to abandon them. The reserve boundaries continued to be maintained and patrolled by uniformed forestry staff.

Forest Reserves were established in what were vacant lands, often in buffer zones between communities. The Government gave preference to protecting major water catchments, unstable terrain and lands that appeared unsuited to arable agriculture. Priority was also given to preserving sufficient blocks of the more valuable forest, which could be efficiently exploited to furnish both domestic and export demands for timber. Moreover, the reserves had an incidental benefit, in that they precipitated a shortage of land for expanding arable cultivation. It might be hoped that this may help to accelerate the change to less wasteful and destructive forms of agriculture.

In Nigeria it became the practice to vest reserve ownership in the Local Authority, who received royalties from timber exploitation within the reserve. Revenues to Central Government came from duties on timber exports, and from company and personal taxes. Where Government participated in the management of a reserve, a proportion of the royalties might be retained to meet this cost. The policing and silvicultural activities within the reserve were usually carried out by Local Government staff, monitored by a Central Government forestry officer. The land covered by the forest reserve was regarded as continuing in the ownership of the communal land owners. Government acted as a trustee to ensure the proper management of the forest estate.

The situation in other developing countries is not necessarily the same. Central Government may create forest reserves 'by decree', and they may be found only as cartographic entities, by lines drawn on a map. Their existence may be disregarded, not only by peasant farmers who swarm up the main extraction roads to obtain access to virgin lands for cultivation, but also by other departments of government, who may annex parts for their own use, or undertake exploitation without Forestry Department permission. The Forestry Department may receive scant support, either from the judiciary or from law enforcement agencies. Thus the whole exercise of creating a national forest estate can become a mockery. More recently in West Africa, the forest reserves have been treated as a source of plunder by unscrupulous individuals. The boundaries have become neglected and overgrown, perhaps partly because this seems an unproductive expenditure of scarce funds, but also sometimes because of the inability or unwillingness of forestry staff to 'go to bush'. Government may license exploitation of the reserve without reference to the local communities, or alter the land use, for example, by reforesting with timber crop plantations, effectively extinguishing the rights within the reserve of the local community, but without consulting or compensating them. Whatever the legal situation, the local population may regard themselves as the natural owners of the land, to do with it as they like. In consequence they may connive with sawmillers to steal trees from the forest, or set fire to it in order to deny to others what they have lost themselves. In Nigeria, the assumption by State Governments of both the executive functions of former local government forestry departments, and the supervisory role formerly imposed by the central government, may have made the situation worse. In effect State Forestry Departments have become their own watchdogs, and both the national interest and the community interest have become neglected.

In Africa, the protection of the forest estate has been helped by other factors. Unlike for example in Brazil or Sarawak, strangers are reluctant to move onto land that does not belong to them, where they have not received permission from those whom they regard as the natural land owners. Onslaught on the reserves has therefore been mainly by encroachment from existing settlements. Unfortunately people do not regard natural forest as an adequate use of land. It is this feeling of waste, as well as economic necessity, that induces people to enter forest reserves to farm. In Nigeria, a response of the Forestry Departments to the risk of appropriation of the reserves has been to abandon natural regeneration work and to use the funds to establish timber crop plantations along access routes and around settlements. This pre-empts the virginity of the soil, thereby removing its greatest attraction for the farmers, who are usually reluctant to walk more than five kilometres from house to farm. However, at the same time, the peasant farmers have been encouraged to cooperate with the Forestry Department, by employing taungya farming in which the farmers clear and cultivate the land with arable crops, which the forestry staff interplant with a tree crop, as described in the previous chapter. When farming is abandoned the trees close canopy to form a highly productive forest, in place of the usual bush fallow which is composed mainly of valueless rubbish. These belts of forest have had the merit of establishing timber crop plantations on a pilot scale, furnishing the quantitative data needed for planting on a major scale. The added expense is justified by the protection afforded to larger blocks of natural forest, and the proximity to lines of communication facilitates the maintenance and exploitation of these plantations. It may be hoped that, because of their greater productivity, they may help to take the pressure for exploitation off the remaining natural forest.

In savanna country, additional legislation made possible the establishment of Communal Forest Areas (CFAs); these were woodlots intended to provide fuelwood, building poles and thatching grass for villages and small towns. CFAs may usually be recognised by the presence of a few trees of exotic species, such as Cassia, Neem or Teak, and by a general air of neglect. They are often on the worst available land near the village, perhaps where over-cultivation has led to hardpan exposed at the soil surface, and much of the area may be more or less incapable of supporting useful tree growth. The unfortunate fact is that what is supposed to be everybody's business tends to be nobody's responsibility. However, if the Forestry Department takes over the management, they may expect to sell whatever wood is produced, in order to recover the cost of establishing and maintaining the plantation. This will be to whoever pays, possibly for the wood to be carried to the urban area. If firewood is in short supply, it may be sold to the friends and relatives of the forestry staff. If firewood is in very short supply, then the members of the Forestry Department may either use it themselves or resell it at a profit to themselves. Meanwhile the rural peasants, who in any event object to paying for firewood, scratch together what firewood they can find by digging up the stumps and roots of felled trees, or by burning cornstalks and dung which are badly needed to manure the farms, or steal the trees from the CFA! The implication is that farmers must provide their own needs for fuelwood and construction wood. This will depend on farm trees and on hedges that belong to individual farmers; but this infringes on land tenure, and on the way land tenure is evolving in the locality, and on concomitant changes in land tenure that are essential for agricultural development.

Similar anomalies may affect shelterbelts, established for the communal good. Where indigenous species are planted, ownership may conflict with legislation on 'Protected trees', the farmer requiring permission of the Forestry Department in order to fell the tree, and perhaps also having to pay a fee into the Government or Local Authority Treasury. The erection of hedges and fences to enable modern farming could also have serious consequences for access by nomadic herdsmen. It is perhaps inevitable that the expansion of arable land, and the progressive establishment of commercial farming, will restrict (and is restricting) the area of land available for nomadic pasturage. This has, to some extent, been catered for by converting some savanna reserves to 'Grazing reserves'; but this is only robbing Peter to pay Paul. For example in northern Nigeria only eight per cent of the total land area is reserved, of which perhaps only half may be suitable for tree crops. This is grossly inadequate to supply the firewood needs of the region. It is to the farmers themselves that one will have to look to grow a substantial proportion of the firewood needed, and perhaps also to produce matchwood and pulpwood, which can also be grown on rotations of a dozen years or less. Land tenure arrangements must make this possible. Moreover, the improvement of pasture, or the raising of fodder crops, are not possible without some form of permanent tenure of land. The expense of breeding to improve meat and milk yields from an existing cattle population cannot be justified without improved feeding to make those increased yields possible, and furthermore from a reduced area of land. In other words land enclosure is particularly necessary for progress in livestock production.

Obstacles to changing land tenure

In 1922, Lugard summarised traditional African land tenure as follows. 'The general principles would seem to be that the assignment of land to the individual is entrusted to tribal or family authorities, in whom is vested the control of common grazing lands and forests; that every individual has a right to a share of the use of the land, and holds it in perpetuity, subject to the performance of tribal obligations, but may not alienate it; that these principles are held more tenaciously by the forest tribes than by those further north ... while pastoral nomads are indifferent to questions of land ownership and value grazing rights only. The forest tribes jealously maintain that all unoccupied land belongs to some community or other, while the northern [i.e. savanna] tribes are less insistent on such claims. All alike recognise the right of the conqueror to dispose of the land. The inevitable tendency to individual ownership is meanwhile constantly asserting itself, with the evolution of the tribe, and is fostered by pressure of population, by foreign example, and by the replacement of annual by permanent crops.'[12] He also wrote, 'It is clear ... that African land tenure is not "communal" in the sense of tenure in common. Its fundamental characteristic seems rather to be an individual tenure of land derived from the common stock at the disposal of the tribe or family. Such a tenure would tend to develop

12 Lugard, F. D. (1922) *op. cit.*

very rapidly into individual ownership, and evidence that this was the case was given by several witnesses – and much more is available.' He continues, 'It is remarkable with what tenacity the native mind holds to the idea of private ownership, or at least of the absolute right to tenure in perpetuity. It was difficult in Lagos to dispose of Crown land to natives on any terminable leases, however favourable the terms.' He quotes Dennett as stating that 'to alienate land by sale was formerly held to be a crime among the Egbas (of Abeokuta), but was then becoming a custom. A deputation of natives had urged on the Ake the view that land was not inalienable, but was the private property of the occupier.'[13] Amongst the traditionally non-Muslim, the family head is the living representative of the ancestors; the land is regarded as belonging to the ancestors for the use of present and future generations. However, in more recent times, this has not prevented families alienating their rights in land if the price was right! It is likely that, as time passes, this conception will become increasingly symbolic, though a piece of communal land, perhaps only a compound, will be retained to represent the lineage.

In Africa at least, there is nothing more likely to bring out man's homicidal tendencies than disputes over land, or over rights to land. Trevelyan, describing England at the time of the Peasants' Revolt, says, 'The peasant profited as much as he lost by the absence of police. The villein farmer striving for freedom, the free workman in constant revolt against the Statute of Labourers, were neither of them in such real subjection to their "betters" as the agricultural labourer in the well-policed countryside of the nineteenth century, when the poor had been deprived of bow and club, and had not yet been armed with the vote. In the fourteenth century, when every man was expected to "take his own part" with stick or fist, with arrow or knife, a union of sturdy villagers was less easily overawed.'[14] I have seen troops of peasant farmers, similarly armed, streaming from the African countryside to challenge a military government, to overwhelm armed soldiers, to let prisoners out of the jail, and to kill unpopular officials – having taken the age-old oath that none shall turn back until they all turn back – and Government, at least for a time, acceding to their demands. I once rescued a villager who had been struck in the back of the head with a battle-axe, wielded by a nomadic herdsman because he insisted on poisoning the pond for fish where the herdsman wanted to water his cattle. I have read reports of villagers fighting with matchets

13 Lugard, F.D. (1922) *op. cit.*
14 Trevelyan, G.M. (1949–52) *Illustrated English Social History*, Longmans Green

over disputed boundaries, and heard of a case where Government sponsored settlers were murdered man, woman and child, by the local tribespeople. Bunting writes, 'There is no need ... to spell out the many difficulties which arise where land suitable for crop production is limited, fully used or controlled by land tenure arrangements which leave both land and the landless idle or insufficiently productive. It is, however, worth reminding ourselves that land tenure is more than a matter for intellectual debate. I know of no circumstances in which significant changes in the control of this critical resource have been effected without either the threat or the reality of the spilling of blood before, during or after the changes. To pass pious resolutions at international gatherings or in parliaments may be a necessary but it is virtually never a sufficient condition.'[15]

The nature of freehold tenure

Purchasing a piece of land 'freehold' is not comparable with acquiring a piece of movable property, with which one can do as one likes. Conversion from communal tenure to freehold tenure is, in fact, a reallotment of the rights of the community. The change strengthens the position of the individual occupier, giving him a more comprehensive possession of personal rights over a piece of land than formerly. Moreover, the community becomes no longer obliged to provide the use of rights in land to all its members, though it may be obliged to provide for their maintenance if they become otherwise destitute. When land is purchased, it is the rights to the land that are being acquired, rather than the land itself – after all, one cannot normally carry it away! Another important change is that 'trespass' becomes both a civil offence and, in certain circumstances, a criminal offence. That is, only the owner, or those who have his permission, have access to his land, and this is reinforced by Government via criminal legislation. There can be exceptions, called 'easements', which permit rights to others over a particular piece of land, such as rights of way, rights to water or rights to drainage. These may be abrogated either by a legislative act or by failure to exercise the right over a sufficient period of years. If land is mortgaged, it is in effect the rights to the land that are mortgaged, rather than the land itself. An important

15 Bunting, A.H. (1981) *Changing perspectives in agriculture in developing countries: an agronomist's view*, Journal of Agricultural Economics 32(3), (pp. 287–302)

consequence of 'freehold' is that these rights become a freely negotiable asset.

The value of the rights differs, according to the use to which the land may be put, and the financial return which this may earn. However, in the price paid for a piece of land, there may be an additional (or subtractional) element called the **development value**. This represents the discounting of a potential change in land use, either for better or for worse. This may depend on infra-structural or political changes in the surrounding country, to which the proprietor of the land may or may not have contributed. It could result from improvements in farming methods (and hence profitability), from changes in relative prices of foodstuffs, or changes in the value of money. If the changes are for the better, it is termed **improvement value**. What people object to is not so much the selling and buying of an individual's rights in land, which has long been practised, nor to an individual raising the value of land by improving it, which has been traditionally recognised, but to windfall earnings accruing from speculative increases in the development value of land. These involve no effort or expense on the part of the land owner, beyond a certain farsightedness or good luck. Nevertheless, speculative land buying can have the merit of ensuring a constant market for land, and of more realistic and stable prices. This helps to make it possible to mortgage the land, and thereby the owner can obtain sufficient money to develop it. Speculation is not objectionable, provided that it does not wilfully remove land from use or enable individuals to hoard land and then to exert the advantages of monopoly, or to make use of privileged information with regard to land development. These undesirable factors can be controlled by fiscal measures, by the laws of inheritance, and by legislation.

Tenure in northern Nigeria

In northern Nigeria it is apparent that compensation for improvements to the former proprietor of the land includes an element in recognition of development value. In other words, the difference of such payments from the purchase of 'freehold' rights in land is not sharply distinct. The real difficulty lies in the stipulation that government permission is required in order to sell, or otherwise alienate land, or to subdivide it between heirs. Of course the original proprietor can assign his rights to another individual, but the second person cannot reassign the rights to a third individual, because the original proprietor retains his proprietorship of the land. Any payment would presumably have to be covert, to avoid the imputation that the land was being bought or sold, or leased for a defined or undefined period.

It cannot be said that the proprietor is the 'owner' of the land, because the owner would appear to be the Government. The only reason that this has not caused confrontation between Government and governed, is that people have continued to implement, and to modify, customary law as though government legislation did not exist. The statutory law may give 'strangers' some security of title, but they could be subjected to harassment or worse in the event of serious land shortages. Land, or rather rights in land, under these laws are not a freely negotiable asset, and hence are not a satisfactory security for mortgage, because it would be impossible to distrain on the land without the willing cooperation of Government and of government bureaucracy. Land cannot therefore be readily used as security to obtain credit to finance its own development. This inevitably inhibits rural development, particularly for people for whom the possession of land is their only asset. It discourages the flow that would otherwise occur of capital, generated in commerce and industry in the towns, into the pockets of rural people.

Lugard wrote, 'The communal system at one end of the scale has its defects, in that it provides no check on the wasteful use of land, by allowing large areas to lie fallow, while valuable forests are destroyed in the search for virgin soil, and in that it checks individual enterprise and industry. Unrestricted freehold tenure at the opposite extreme has its failing, in that there is no check on the mortgage or sale of his land by the peasant holder who, as has happened in India, infallibly becomes the prey of the moneylender, and a tenant or serf on the lands he once owned.'[16]

It is clear that Lugard intended land tenure legislation to protect the peasant from exploitation, possibly particularly by Europeans (at that time). However it is also clear that he did not expect these arrangements to be permanent. He wrote, 'Outside these districts [i.e. big coast cities where the concept of freehold land was already introduced] it would seem to be desirable to enforce registration . . . of all holdings not already registered, in which individual ownership is claimed, whether by aliens or by natives. The remainder of the land should be declared to be subject to native tribal tenure and, except to Government for public purposes, alienable only in accordance with

16 Lugard, F.D. (1919) *op. cit.*

the principles of native law – e.g. the freehold should not be granted. If at any time it were desired to extend the application of English law to any transfer of land in such areas, the consent of the Governor would be required, in order to ensure that the grantor fully understood the transaction . . . This procedure would not arrest natural evolution, or perpetuate by legislation a system which is in course of development, but it would ensure that native tenure could only be modified to accord with changing circumstances, after due consideration, and with the assent of the Government . . . By such means the Government, through the District Officer, could assist the native chiefs in their dealings with Europeans . . . It is the first duty of a government to watch over the interests of the governed, and if there is one section of these which claims a special share of consideration and thought, it is the one which, by reason of poverty and ignorance, is least able to take care of itself.'[17]

Conclusions

The natural progression, in Africa, from degrees of communal land tenure towards individual ownership is absolutely clear. This development has been assisted by the mutability of customary law, which has responded successfully to changing social and economic conditions. It explains the variations in customary land law, even between adjacent communities. Unfortunately the governments of some countries have tended to be more reactionary or doctrinaire than the traditional rulers. They appear to restrain, rather than to encourage, the evolution of land tenure. Perhaps this is more from a fear of the inconvenient problems that go with land tenure reform than from any mistaken belief that the ancients knew better how to solve modern predicaments. Sometimes this retrogression is in the name of 'Socialism', though the results do not suggest that it is for the benefit either of the people or of the State.

'*God helps those who help themselves.*' I am not sure if this is good theology, but it is good sense. People will respond to the challenge themselves, of their own accord, but it is necessary to provide the legal framework within which change can occur. This should be with a view to arriving at a consistent form of land tenure nationwide, but allowing the different parts of the country to move at their own pace in their own way. The new form of tenure must be capable of

17 Lugard, F.D. (1922) *op. cit.*

responding to changes in economic circumstances, so that land development will not be impeded by anachronistic arrangements, nor by bureaucratic inadequacy. For example, it should be the duty of government to register a land transfer, rather than the duty of a citizen to succeed in registering a transfer before its legality is recognised. Land laws, and the rules and regulations governing transfers of rights in land, have to be subject to modification in the courts and by statutory legislation, otherwise they cannot be sufficiently adaptable to meet changing circumstances. Moreover, it is necessary to perceive and comprehend the other consequences of land reform, and to take whatever measures are possible to mitigate the ensuing repercussions. It is apparent that alternative systems of agriculture must be found that will use the land more efficiently, and that these may not be suited to the scale of peasant agriculture. In consequence, land could no longer be used as a safety net for the mass of the population. There has to be a move towards urbanisation and industrialisation, and at the same time towards more capitalised and more productive forms of agriculture to support growing non-agricultural sectors. There is no choice.

The next chapter will attempt to explain the nature of agrarian reform, by looking at the European experience. In Africa the slogan 'Green Revolution' has been preferred to the term 'Agrarian Revolution'; but perhaps this is to avoid thinking about the necessity for land tenure reform, and the upheaval that must ensue. However, in consequence, we may be building a house without foundations, that keeps tumbling down whatever may be done to shore it up. There comes to mind the parable of the '*wise man, who built his house upon a rock: and the rain descended, and the floods came, and the winds blew, and beat upon that house; and it fell not: for it was founded upon a rock*.' And there is also the '*foolish man, who built his house upon the sand and the rain descended, and the floods came, and the winds blew, and beat upon that house; and it fell: and great was the fall of it*.'[18]

18 *The Holy Bible, Matthew* chapter 7, verses 24–27

5 Agrarian revolution

> *'All our properties, all our laws, all our manners, all our minds changed. This, which I noticed, took place within forty, and most of it within ten years.'*
>
> *William Cobbett*

> *'New and better systems supersede old systems to the advantage of the nation and of those to be employed, but the progress from worse to better involves hardship to those, and they may be a larger number, whose ideas and powers are adapted to the methods of the past.'*
>
> *E.C.K. Gonner*

In traditional societies, farming is directed at providing nearly all the needs of isolated communities but as development proceeds the need is to feed a nation. 'Undertakings in husbandry were generally small, calculated rather to be a means of subsistence to particular families than a source of wealth to the public.'[1] Agricultural productivity expands to feed burgeoning populations, by opening new land for cultivation. Eventually good land comes in short supply and less favourable land is brought into use. As the resource becomes overexploited, agricultural productivity begins to fall. Despite improving agricultural wages 'Labour is scarce . . . young ones refuse to learn farming . . . They seemed to be left with the old men, the cripples and the dullards.' A farmer complains, 'I may say that this day five young men have left the next village for London, and we are so short of labour that we cannot get our work done, but it is looked upon as a sort of disgrace to work upon the land.'[2] Young women also desert the rural areas. 'The life of the small farmer's wife is a life of continual labour indoors and out of doors. Many a young Irish girl must have looked on the wrinkled face and bent back and rheumatic limbs of her mother, and grown maddened in a sudden passion at the thought that her own fresh young life might end like this, and must

1 Horner, quoted by Gonner, E.C.K. (1912) *Common land and enclosure*, Macmillan

2 Haggard, H. Rider (1902) *Rural England, being an account of agricultural and social researches carried out in the years 1901 and 1902* (2 vols) Longmans, Green & Co (pp. 584 and 623)

have made up her mind that life on an Irish farm was no life for her . . . I am certain that the migration of women from the land will continue and even increase. It is a period of awakening intensified self-consciousness for women. Women's rights are everywhere being discussed. They are comparing their share of the world with man's, and growing more and more dissatisfied.'[3]

Ultimately major reform is inescapable; but in a traditional community, farming system and social system are meshed together –change one and the other is disrupted. That is why, in England, land reform was called the 'Agrarian revolution'. It is worth reviewing the nature and consequences of this revolution, in relation to what is happening in parts of Africa, although one realises that there can be no exact transference across space and time. Samuel Johnson wrote, 'If no use is made of the labour of past ages, the world must remain always in the infancy of knowledge. The discoveries of every man must terminate in his own advantage, and the studies of every age be employed on questions which the past generation had discussed and determined. We may with as little reproach borrow science as manufactures from our ancestors; and it is as rational to live in caves 'til our hands have erected a palace, as to reject all knowledge of architecture, which our understandings will not supply.'[4]

We must remind ourselves how relatively recent the transformation was in Europe, and that England was the first to experience this change, which took place after Dr Samuel Johnson wrote the paragraph quoted. In the English village where I spent my childhood, the conversion from communal land tenure to freehold tenure took place in 1829. Before the Second World War, when tractors had only just begun to replace horses, ploughing was not so deep, and the patterns of the common field could still be seen deeply scored in the land. Except for the use of draught animals, the English communal farming system compared with the most primitive systems surviving in Africa. Three common fields surrounded the village. One field carried crops of wheat, oats or barley, the second field was peas and beans, and the third field was left fallow. The fields were rotated in turn. Also incorporated in the fields were smaller patches of other crops such as flax (for linen cloth manufacture) or potatoes. Land retains its fertility much better in temperate climates, as explained in Chapter 3,

3 Russell, G.W. (1912) *Co-operation and nationality – a guide for rural reformers from this to the next generation*, Maunsell & Co, Dublin, reissued in 1982 by the Irish Academic Press

4 Johnson, Samuel (1751) Essay from *The Rambler*, No 154, of Saturday September 7, 1751

and a fallow of one year in three was sufficient to maintain its fertility. The fields were unfenced, and were ploughed for the benefit of all, for the village might only be able to afford one or two ploughs. The fields were divided into strips, each measuring about one acre, which was the area a team of oxen could plough in a day. Each peasant cultivated several strips scattered through the fields. With growth of population, fragmentation of holdings occurred through inheritance, and some persons might have only part of a strip in each field. The **Lord of the Manor** sat in court and appealed to juries of commoners against men who neglected their duties, and who could be fined. The lord usually had his own separate estate, called the **demesne**, though sometimes he shared in the common fields.

Formerly, the villagers had to spend a certain numbers of days in the year working on the lord's land, but this had already been discontinued and replaced by a monetary rent, and he hired labourers for his work. The manorial courts were later replaced by vestry meetings of commoners which regulated the management of the common fields.[5]

Agrarian reform in England

Pasture for cattle was also unfenced and in common. Meadows near the stream, which were too wet for arable farming, were retained as permanent grazing; also the cattle could be pastured on the fallow and in the fields after the harvest; at other times they could roam in the *waste* (or 'bush' in African parlance), as the unallocated land that surrounded the village fields was called. According to the number of strips he held in the fields, the peasant obtained a *stint*, that is a certain number of cattle that he was allowed to keep. The village herd was watched over by a cowherd, the pigs by a swineherd, and the sheep by a shepherd, who made sure that the livestock did not damage the crops. There were also people called *cottagers*, who were squatters on the waste but had no rights in the common field, and had a house surrounded by a small garden, and let their livestock run on the waste. In my village, the waste was part of the Charnwood Forest; but, by that time, felling of the timber for house construction and firewood,

5 The Manor was a territorial unit, consisting of the lord's demesne and of the lands from whose holders he could exact certain dues. The commoners were, of course, the persons farming in common. The manor often coincided with a parish, that is an area having its own church and clergyman. The vestry was a room or building attached to the church, and hence a meeting of parishioners in the vestry for parochial business.

the annual burning to promote a flush of pasture, and prevention of tree regeneration due to grazing, had reduced it to a barren moor, carrying a low stock of animals of poor quality. At night the livestock could be folded in the fallow field, so that their dung would manure the land.

The peasants could transfer the proprietorship of their holdings, or part of them, to other people by sale, bequest or marriage settlement, together with the associated rights of common (i.e. including pasturage). If strips were abandoned, or fell out of cultivation due to a lack of heirs, they reverted to the lord of the manor, who might cultivate them himself or find another person to take them over. Neglected weedy strips were not desirable in the common field. The peasants normally left their holdings to an eldest son, in order to keep it intact, and to add to the area if possible. (Perhaps a reason why England became an aggressive imperial power, despite the lukewarm enthusiasm of its governments, was because, if land was short, the younger sons had either to earn their livings as hired labourers, or squat on the waste – or seek their fortunes elsewhere.) People might also appropriate patches of land, excised from the waste and fenced, which if they were held for a sufficient period of years could be regarded as freehold land. The Hammonds write, 'Whatever the pressure outside and whatever the bickerings within . . . the common field system formed a world in which the villagers lived their own lives and cultivated the soil on a basis of independence.'[6]

Land was regarded as something that supported a community. Agriculture was a way of life, and fixed the cycle of seasonal activities: in the field, in the home, in the church, round which the social life of the village revolved. Food was grown primarily to feed the village itself, and the farming system was directed at producing a guaranteed minimum yield to meet all the needs of the community. In a good year there would be a surplus to ensure adequacy, which could either be sold or stored against future shortage. The system was stable and resistant to change, because this could be achieved only by agreement between the participants. It gave the people free time to engage themselves to some extent in manufacture, both for their own needs and perhaps for sale; but latterly in an industrial age this was no longer necessary. The community supervised and controlled the performance and welfare of every individual.

It was said of traditional agriculture, 'Although the common field husbandry does not make the land better, it keeps it from

6 Hammond, J.L. & B. (1911) *The village labourer 1760–1832*, Longmans Green and Co

becoming much worse.'[7] This could also be said of peasant farming in Africa. Traditional farming practices *do* respond to market forces, but only to the extent that the whole socio-agronomic system is not overthrown. In England, when between two-thirds and a half the working population still lived mainly by farming, agricultural productivity did not rise fast enough to support growing populations adequately. The reasons were that traditional farming systems could not be made profitable enough to attract capital investment, and that the new farming systems required more integrated use of land. These depended, in turn, on consolidation of fragmented land ownership, and on disencumbering land of traditional communal rights in order that it might pass to its most economic use. This could not be achieved without drastic changes from traditional land tenure.

The declining suitability of the traditional common field system was demonstrated by the inadequate growth of agricultural production at a time of upsurge in population. The expanding urban areas, and rising standards of living, meant a growth in demand particularly for meat, milk, wool and leather. The rural people could supplement their diet with rabbits, pigeons and wild game ('bush meat' in West African parlance) but this was not so easy for town dwellers; nor could the supply have been sufficient. The difficulties were due to an extension in the area of cultivated arable land, and hence disappearance of the waste. This reduced the land available for pasturage, and also for the forest produce that it furnished – especially important for the poorer members of the community. The disadvantages of the open field were clearest in relation to animal husbandry. The indiscriminate herding together of cattle and sheep of all owners made scientific breeding impossible, and increased the danger from infectious diseases. Though *stinting*, or regulation of numbers of livestock, might be practised, there was a tendency for each individual to try to extort the maximum for his own benefit, and hence for overgrazing. This sometimes reduced the common pasture to near dereliction, as it had done in my own district. Moreover, the lack of feed during the winter made it necessary to kill off a good proportion of the animals in the autumn, and to preserve the meat with salt. Indeed it was these circumstances that propagated European trade with tropical countries, in order to obtain the spices that could make badly preserved meat edible. A result was that the number of animals that could be kept was limited by the number that could be maintained during the winter. It was vital to make agriculture more productive,

7 Quoted by Gonner, E.C.K. (1912) *op. cit.*

and thereby to make farming more profitable in order to justify capital investment in the land. This meant developing systems of permanent agriculture, in effect by learning to use the fallow, which in England was pasture.

A solution was made possible by quantifying individual rights in proportions of land area and dividing up the land among the communal proprietors, and taking the opportunity to consolidate and fence the previously fragmented holdings of each individual. In some parishes this had already occurred by agreement, but it was essential that the process should have the backing of law, so that a minority of recalcitrant proprietors could not refuse to cooperate, or afterwards recant and thereby wreck the whole scheme, but could be compelled to go along with the majority. Eventually a formal procedure was established, known as 'Parliamentary enclosure'. Parliamentary Standing Orders required that the proprietors of at least *two-thirds of the area* of common land should have endorsed the change before enclosure could begin. The procedure began with a public meeting, called by those anxious for change; a draft petition, signed by witnesses, was prepared for submission to Parliament to show the degree of consent. The Draft Bill was drawn up, and reviewed by a Select Committee of Parliament, before whom the petitioners had to appear or be represented, and by whom any counter petitions could be heard. The Committee then reported to Parliament, stating that Parliamentary Standing Orders had been complied with, and confirming that the petition had at least the required degree of consent. In order to satisfy objections, amendments might be recommended for insertion into the Bill. The Bill then received first and second readings, followed by Royal Assent, or was referred back to the Committee where it could be amended or dropped. Commissioners, usually three in number, were selected by the petitioners and appointed by the Private Act, which might impose other conditions to regulate the enclosure, such as:

- the proportions of land area to be awarded to extinguish the 'rights to the soil' of the lord of the manor, or in lieu of tithes;[8]
- perhaps to regulate the manner in which the cottagers were to be compensated, though strictly speaking they might have no rights to the common at all;
- to compensate those inequitably losing improvements or being disturbed by the change over;
- making provision for the management of the area during the interval between the Act and the Award, in order to minimise disturbance.

The commissioners were men of standing and integrity, who were bound by oath to administer the work with justice. They were prohibited from having any interest in the area where they were working, and from owning any land there for a least five years after the Award. Probably the best commissioners were practical men with knowledge of farming and surveying; some had legal experience. Generally at least one commissioner was from nearby, with a good knowledge of local circumstances. They were usually assisted by a clerk, a valuer and a surveyor, though sometimes they might carry out some of these operations themselves. They gained experience from being employed in enclosure after enclosure by different parishes.

Thus the force of law was given to a process that had begun to occur naturally, for it was important to ensure that an enclosure was completed and that its validity should not be challenged afterwards. It was accomplished over the country when each community undertook to change and could accommodate local circumstances. The village where my father was born was enclosed in 1765, more than sixty years earlier than the village four miles away, where we now live. My father's village had no lord of the manor, for it had belonged to Leicester Abbey, and on the Dissolution of the monasteries the demesne had been seized by the King, and broken up and sold to the local peasants. It was possible to receive a higher price in this way, than by selling to a single purchaser. In this village, the period of the settlement from Act to Award took only eighteen months. In our adjacent village it took twenty years, because the Charnwood Forest was shared among seven parishes. The bulk of Parliamentary enclosures took place during a period of one hundred years, between 1750 and 1850. An enclosure act might refer to the common field, or to the common pasture, or to the waste; it might include any or all of these. Thus the subdivision of the land might be in one bite, or in several bites. It was written at the time, 'The power granted by most of

8 The 'tithes' were originally a contribution of a tenth of the produce of the soil for the upkeep of the Church, and for charity, but might have passed into lay hands since the Dissolution of the Monasteries in the sixteenth century. The tithes were commonly exchanged for a fifth of the open field, a seventh of the old enclosure and a ninth of the common waste. In lieu of rights to the soil, the lord of the manor might receive a grant of between a sixteenth and a twentieth part of the common or wasteland, i.e. that part not under individual proprietorship.

The system of tithes had the economic disadvantage that they were a proportionate decrement of crop yields, and therefore a disincentive to production. As a social tax, they were replaced by the 'poor rate' which was a flat charge based on the rentable value of the land, i.e. related to potential productivity, and therefore an incentive to production.

the Acts to Commissioners is an extra-ordinary circumstance . . . They are a sort of despotic monarch into whose hands the property of the parish is invested, to recast and redistribute at their pleasure among the proprietors, and in many cases without appeal. But this great confidence is necessary, for such is the complexity and variety of interests to be examined and estimated, that if more cautious methods were resorted to, it would be impossible to effect the purpose.'[9]

The initial duty of the commissioners was to carry out a survey and valuation of the area to be enclosed. Besides consolidating holdings, it was essential for them to plan, lay out and realign roads in order to provide access. Sometimes provision was made for drainage, and on at least one occasion for irrigation. Each allotment was subdivided into compact fields, and in order to minimise the expense of fencing, the boundaries were in straight lines approximately at right angles to one another, which the owner had to fence within a certain time, usually by means of ditch and hedge. The average field size was about ten acres (four hectares) which was found to be about the optimum for stock control. The commissioners faced the dilemma between consolidation of holdings and economy of fencing on the one hand, and the farmer's requirement for varying types of land on the other. Field boundaries were drawn up so that each field was as homogeneous as possible in terms of soil and productivity. Moreover, it was realised that smallholders did not possess capital to improve their land, nor an adequate income from their holdings, and that they should therefore receive the better land near the village, so that they could supplement their incomes by following other trades, or by working as labourers on the bigger farms. The larger land owners could be given their holdings nearer the margins of the parish, as they could afford to build a new house and farm buildings on their land. The treatment of former enclosures varied according to circumstances. Normally, large consolidated blocks were left alone, but to minimise disturbance the commissioners might well place any new allotment adjoining the proprietor's old enclosures. Costs were considerable, and were paid either by the proprietors individually and/or by sale of a part of the common lands. Where proprietors were unable to pay the charges, or perhaps to carry out the fencing, the commissioners completed the work, and then put the land out to rent until the debt was discharged.

The expenses could have been considerably reduced if there had

9 Young, Arthur (1808), quoted by Yelling, J.A. (1977) *Common field and enclosure in England, 1450–1850*

been a General Act of Enclosure, laying down the rules and regulations of enclosure and the method by which it ought to be accomplished, and by appointing official commissioners. This eventually happened, but almost too late to have much influence on the progress of events. Many Members of Parliament were lawyers, and had a vested interest in the *status quo*. However the reforms successfully boosted agricultural production and nourished the industrial revolution of succeeding decades. Yields quickly rose by a third, and profitability almost doubled, not so much due to any immediate change in farming methods, but because of greater efficiency arising from consolidated holdings, and from land becoming put to optimum use depending on soil, climate and local market forces. Dudley Stamp wrote, 'It laid the foundations for all real development in farming practice.'[10]

The integration of agriculture

Improvements depended on a technical breakthrough: the amalgamation of arable cropping, animal husbandry and forestry in ways impossible under traditional systems. It necessitated learning to use more fully the fallow, which occupied a third of the cultivable land. The introduction of the turnip (a root crop), and the intentional sowing of pasture with improved grasses and a legume (clover) as a part of crop rotations, produced deliberate fodder crops and made it possible to maintain large numbers of animals through the winter. This augmented the provision of meat and animal products to the towns, and at the same time the increased supply of manure gave richer crops and enhanced grain production. The cattle were kept in stalls at night, and the bedding straw, the dung and even the urine and washings from the milksheds were returned to the fields. The fields were drained, often with underground tile or stone drains, though in regions without stones, brushwood, taken from the hedges or from the waste, might be used. The ditches were cleaned out periodically, and this debris, together with the sludge from the ponds, also helped to fertilise the fields. Improved pasture had to be justified by improved livestock, and vice versa. Robert Bakewell, who lived not more than ten miles (16 km) from my village, was the first to begin to breed cattle to increase their bodyweight, particularly on the more valuable parts of the carcase. He also improved milk yield of cows, and

10 Dudley Stamp, L. (1957) *Britain's structure and scenery*, New Naturalist Series, Collins

the wool yield of sheep. The seed drill was introduced, which gave a better distribution and germination of seed than broadcast sowing by hand, and yields could be three times higher from the same amount of seed sown.

The hedgerows, besides preventing trespass of cattle, were an important source of fruits and nuts, and also of smallwood for making hurdles and for stakes and fuel. Trees, such as ash, elm and oak were also planted in the hedges as sources of estate timber to provide rafters, furniture and coffins, as well as gates, cart shafts and bottoms, and tool handles etc. The trees have now reached maturity and, in England, it has been estimated that a third of the total stock of timber stood in the hedgerows – that is before the recent scourge of Dutch elm disease. This compensated to some degree for the country being very inadequately wooded, with less than ten per cent of its total area dedicated to forestry. The irregular structure of the hedges is also an effective windbreak, and the ditches contributed to field drainage. It has to be admitted that, today, the hedgerows are a very under-utilised resource. In fact, their use as a source of fuel in rural areas was curtailed by the construction of the railways, and the improvement in communications. This permitted coal to be carried cheaply to nearly all parts of the country, even to the villages. The interest of farmers in managing their hedgerows has also been diminished by a century of cheap timber from overseas; for the past thirty years the neglect of mixed farming, and therefore in some localities the absence of cattle, has reduced the necessity for hedges. Machine cutting of hedges, or their incineration when burning stubble, or even bulldozing them away to enlarge fields for cereal crops is destroying existing hedges and preventing regeneration of the trees. The wind is again beginning to blow across the countryside with noticeably increased strength.

Besides hedges, occasional copses (wood lots) remained as cover for wildlife: in particular rabbits, hares, pigeons, partridges and pheasants. Because of the game laws, and the fencing and conversion of land to individual ownership, access was prevented to the general public. This effectively protected wildlife, and though evidently farmers could never entirely either retain them on their own land, nor banish them to someone else's, they did obtain a measure of control. Sporting land owners, who had an interest in hunting, employed gamekeepers to guard and control the game on their land. Foxes were generally protected for the esoteric pleasure of fox hunting, though farmers might regard this predator with mixed feelings in view of the havoc that it could wreak if it got in amongst their poultry. When cereals were harvested, the reapers with their scythes (and later the

Typical English countryside; note the hedgerows

binding machine) passed round and round the field until the game were concentrated in a central patch of uncut corn, and were clubbed or shot as they fled from the remaining cover across the surrounding stubble. In winter, when the ground was too wet or too frozen to work, the farmer and his men went rabbiting with ferrets. Small doe ferrets were sent into the holes to cause the rabbits to bolt from their burrows and to be netted or shot as they emerged from the holes; or caused to 'tuck', when they blocked the burrow with their backsides to prevent the ferret reaching their throats. Then, when all was quiet, buck ferrets were put down on collar and line, to trace the tucked rabbits, which were dug out and killed. As well as providing meat, this could be an important source of income for a farmer's son, who might receive no wage from his father.[11] He might also be given a runt pig, or orphaned calf or lamb to raise and to sell. His mother obtained money by selling milk locally, and by being in charge of the poultry.

11 Some of the industrial farmers regarded rabbits as vermin, and in the 1950s the sport of rabbiting was virtually eliminated by the deliberate and clandestine introduction of the epidemic disease of rabbits, Myxomatosis, which wiped out whole populations of rabbits.

The new farming systems were developed by larger land owners, who had money to invest, often accumulated in commerce and industry, and who were passionately interested in progressive farming. Some increased the value of their estates ten times during a working life. The discoveries on which the new methods were based began in the second half of the eighteenth century, but it was only during the first half of the nineteenth century that they spread throughout farming in general. Peasant farmers might watch a neighbouring 'gentleman' farmer make his fortune for a decade or more, before one followed suit and the rest followed him. By 1830 the use of what came to be known as **mixed farming** had become widespread; that is, the integration of arable crops, livestock, forestry and to some extent wildlife. This inflated the biological productivity of the land, and hence increased production for human consumption. It depended on growing fodder crops to feed animals, which generated manure, making it possible to keep the land in constant production, and on rotating crops to control weeds and diseases. This was coordinated particularly with livestock breeding; though breeding of staple crops only became significant in the twentieth century. The old common field system, while suited to a period of local isolation, had one inevitable defect: it prevented land from being put to the use to which it was best suited. At the same time it inhibited capital investment and stifled individual initiative. Commercial considerations could be accommodated in the common field only to the extent to which they were compatible with the survival of the system. It could not respond to new circumstances because communal decision-making was always hampered by the innate conservatism of some members or by irreconcilable conflicts of interest. The new farming systems were intended to feed industrial and urban societies, whereas the old systems were directed at feeding an essentially rural society in which the majority of people were farming.

It may be asked why peasant farmers were anxious to enclose their land, since they did not immediately adopt the new methods of farming, but simply continued with old ways on their individual holdings. The reasons were that the consolidation of land holdings improved agricultural efficiency and caused an immediate rise in land values by between two and four times. More important, the farmers could now obtain capital by mortgaging their land. Thus rural people obtained an advantage over the urban population. They could raise money to build houses to occupy or for rent, and the village population was able to increase. They could also borrow money to acquire more land to enlarge their farms. If the availability of land was restricted, because of a larger population, they could look around for other avenues for their enterprise. This turned itself towards industry; my father's village became engaged in the manufacture of boots and shoes and hosiery (stockings). There was also a tannery, a cardboard box factory, a printer and two bakeries in the village when I was a child, and the watermill was still operating to grind flour. This ability to raise capital effectively transferred wealth from commerce in the towns to agricultural and industrial production. It also provided employment in rural areas. The consequence was that, in England, the industrial revolution began in the villages, not in the towns.

The social repercussions

There were social repercussions. Land ownership patterns did not change much for a generation after enclosure (15 to 25 years). The next phase was the accumulation of land holdings by larger or more efficient owners. Small land owners had neither the capital to develop their land, nor the prospect of adequate remuneration for their work. However it was their children, rather than themselves, who abandoned farming. The new system of land tenure was flexible, and permitted farm size to adjust to the optimum for various crops and to respond to changes in economic circumstances. The land tended to become amassed in the hands of large land owners. The new wealth, created by commerce and industry, ensured ready purchasers for land. They were 'quite different men, men of ideas'. Gonner wrote, 'New and better systems supersede old systems to the advantage of the nation and of those to be employed, but the progress from worse to better involves hardship to those, and they may be a larger number, whose ideas and powers are adapted to the methods of the past.'[12] In traditional society about ninety per cent of the working population had been engaged in agriculture, and perhaps two-thirds when Parliamentary enclosure began; but by the time of its conclusion this had fallen to about a third. Inadequate farmers, for whatever reasons, were pushed off the land by economic forces. Because the number of available jobs was dwindling, agricultural workers became (and still are) relatively badly paid. The displaced peasants needed other sources of earning a living, and these were provided mainly in manufacturing industry.

Cobbett wrote, 'All our properties, all our laws, all our manners, all our minds changed. This, which I noticed, took place within forty,

12 Gonner, E.C.K. (1912) *op. cit.*

and, most of it within ten years.'[13] Hoskins writes that enclosure 'was one of the most momentous events in the long history of the village. It transformed the physical landscape of the parish within a few years, altering its farming almost beyond recognition, and changed the entire culture and habits of the peasant community.'[14] Thirsk writes, 'After enclosure every man could fence his own piece of territory and warn his neighbours off, the discipline of sharing things fairly was relaxed, and every household became an island to itself. This was the great revolution in men's lives, greater than all the economic changes following enclosure. But on the other hand the necessary interdependence and common governance of the common field was not an ennobling experience in "pure democracy" but a mass of petty restrictions and a potent source of squabbles and individual friction. There was a release of energy and initiative by enclosure, which enabled individuals to fulfil their personal potential, as well as improving their own lot and that of their fellow citizens.'[15]

It was feared that releasing land owners from the guidance of custom would allow them to attempt to gain too much from the land and to damage the soil, but this did not happen to any great extent. It is interesting also to note that, not only were the different branches of agriculture reintegrated together, but so were the roles of husband, wife and children: each had a function or functions within the new system, and each his or her own reward, though not necessarily entirely commensurate with the effort expended.

In 1954 Dumont wrote, 'The three-field system, essentially an arable system, a form of which survives in north-eastern France to this day, was abandoned in England during the course of the seventeenth and eighteenth centuries.[16] The pendulum swung away from communal enterprise, the animals belonging to the village were no longer herded together on the stubble and bare fallow, the people left their nucleated settlements and re-established themselves in the midst of their farmland in small dispersed communities . . . The characteristic features of this change were the enclosures, and from this time onward

13 Cobbett, William (1830) *Rural rides,* reprinted by Dent Dutton, 1925
14 Hoskins, W.G. (1957) *The Midland peasant,* Macmillan
15 Thirsk, J. (1957) *English peasant farming,* OUP
16 Dumont, Réné (1957) *Types of rural economy,* Methuen
(translation (1954) *Economie agricole dans le monde,* Paris)
Surely the period should be the eighteenth and nineteenth centuries, although antecedents of the change extended back into the sixteenth century. Also, the communal spirit exists to some extent in my own village up to the present day (1985) even though it is now a dormitory village to the City of Leicester.

Typical European countryside with open fields

every field had its hedgerow or wall. This enabled each farmer to graze his animals separately without having to employ a herdsman; this, unfortunately, is still not true in many parts of France. The grouping of the fields around each farmhouse and the virtual integration which resulted therefrom made for appreciable economies in the costs of production. At the same time the very small enterprise tended to disappear. The enclosures may have had certain painful consequences, but they created a class of independent farmers, which quickly emancipated itself from communal obligations and, with a spirit vastly different from that of the European peasantry as a whole, began to improve its farming methods. There was none of the "land hunger" which was absorbing most of the French peasants' savings at the time, and the farmer did not hesitate to devote all the money he could spare to purchasing equipment, to land improvement, and to selective breeding. Even today [i.e. in 1954] 65 per cent of English farms, covering 70 per cent of the agricultural land, are worked by tenants; in France those proportions are almost reversed, 70 per cent of farmers owning their land, which accounts for 60 per cent of the agricultural area. The prevailing attitude in England was that to buy a farm large enough to run efficiently was too costly and left too little capital for equipment and improvements ... There is thus no equivalent to the intractable and conservative peasantry with which France is still afflicted, and in its absence there has emerged, more clearly than in most countries on the Continent or even in north America, a social grouping into "farmers" and farmworkers. Whereas in France, as in the USA, less than a fifth of the agricultural labour force consists of wage-earning employees, the proportion rises to nearly two-thirds in England [in 1954]. Furthermore, on a medium-sized French or American farm employing two or three men the farmer often does more hard physical work than his employees, but his English equivalent quite often refrains from manual labour except at the busiest time of year. He certainly keeps abreast of technical developments more effectively than the peasants of less advanced countries but, all the same, only certain small areas in the south and east reach the standard set by the Netherlands and other countries of extremely intensive agriculture, where the most progressive farmers do not scorn to undertake heavy manual work. England is ninth [i.e. in 1954] among the countries of Europe for its gross yields per acre.'

The conversion from communal tenure to individual tenure was not achieved without human cost. More efficient farming meant that less labour could produce more food. This augmented fears that a result would be depopulation of the countryside. In fact this scarcely happened: depopulation proceeded to a greater degree in parishes that enclosed late than in those that enclosed early. The population continued to multiply, but the surplus had to look for employment in commerce and industry, and in the towns – or to emigrate to new lands overseas. Paupers could no longer find a piece of 'waste' from which to scratch a subsistence, because all the land had passed into private ownership. Provision for the destitute was obtained from the 'poor rate' levied on farmers, each parish being responsible for those born there, or resident there for a certain period. Certainly this checked the mobility of labour, because people were afraid of becoming separated from the only source that might give them relief, and the parishes drove out strangers to avoid the risk of becoming responsible for them. As the number of landless poor increased, the cost of maintaining them began to impoverish agriculture, and enmity was generated between farmers and the labouring class. Rick burning and molestation of farmers by desperate agricultural labourers gave rise to fears of revolution by the proletariat, such as had already occurred in France. This was restrained in England by the legal enforcement of providing subsistence for the destitute; but the system became inadequate as the strains increased. Because of these anxieties, riots were fiercely suppressed by the authorities.

The Government made two critical mistakes. Firstly the poor rates were assessed by the parish according to the rentable value of land, and in addition there was a land tax payable to Government. These had the merit of forcing people to use the land they owned. However the impositions bore most heavily on the small farmer, whose standing charges were so much greater a part of his income than for the large farmer, though he was not necessarily a less productive user of land. He also had less option of selling or mortgaging some of his land, in order to realise capital to tide himself over temporary difficulties. Secondly, by means of the infamous 'Speenhamland system', wages were subsidised out of the rates if earnings fell below subsistence level. This helped to pauperise the labouring classes, and by impoverishing agriculture also impoverished industry. What farm-workers could not buy, the factory workers could not be paid to produce. It would have been better to have had a sliding scale for taxes on land, which would have borne proportionately more heavily on the larger land owner, both to discourage excessive land holdings and to encourage the smallholder. Furthermore it would have been better to enforce a statutory minimum wage, in order to ensure a living wage and to stimulate the national economy, as had been done previously and is done now.

The suggestion has been made that Parliamentary enclosure did not meet the implacable resistence offered by peasants to enclosure in the past, because the power of the governing class had become insuperable. I suspect that it was rather because all had seen the writing on the wall. The poorer peasants were in a deteriorating situation which was, in any event, becoming untenable. Probably most realised that the welfare of future generations depended on reform. Nevertheless it was a plunge into the dark, from which few could guess the consequences. Perhaps we see something like it in Britain today, when an attempt at industrial reform is causing massive unemployment, to which the victims grudgingly acquiesce, in the hope that from their difficulties something better may emerge. Like conscript soldiers going to war, they know they may be sacrificed in the common cause, but are aware that for the community as a whole the alternatives may be worse. Both in agriculture and in manufacture, changes occurred that promoted the emergence of two classes: a small class of employers and a very much larger class of employed. Nevertheless, for most people, this was simply another form of communal control, which outside fixed working hours was much less constraining than that of traditional society, and in which young men might have as great access to wealth and status as the old. Peasant society provided no such opportunities.

Farming boomed during the early years of the nineteenth century, partly because the Napoleonic Wars isolated British commodity markets. For a time England became a grain exporter, until growing urban populations swallowed up the excess. Responding to the cries of the urban masses, to which were joined those of the manufacturers who wanted cheap labour, in 1846 the Government repealed the Corn Laws, stripping protection from home agriculture and allowing cheap grain to be imported into Britain from north America. This was facilitated by the growth of international shipping and of railways. The farmers reacted by moving away from cereal production towards fodder crops and livestock, and after a period of hardship entered a more affluent period during the 1850s and 1860s. But in the mid 1870s, farmers' difficulties were compounded by the advent of refrigerated ships, which enabled the import of cheap meat from overseas. English agriculture was unable to meet these dual challenges, and teetered on the edge of insolvency. It became capable of producing only twenty per cent of the nation's food requirements, a situation that continued until the beginning of the Second World War in 1939. Land accumulated in the hands of too small a number of land owners, and ultimately of an increasing proportion of absentee landlords, who used the land as sporting estates, which they subsidised from their other activities in commerce, in industry or in finance.

In the few places it was tried, retaining some parish land as allotments to be let at low rents enabled landless families to retain their independence, and prevented them from becoming public charges. However the numbers of needy people would eventually have exceeded the capacity of any such expedient, and industrialisation was the only ultimate solution. The burden ought to have been transferred much sooner to industry, where the owners of capital were accumulating vast riches. Eventually this happened with the progressive establishment of a Welfare State.

Agricultural cooperation in Scandinavia

While English agriculture languished, English farmers could see the success of the Danes selling in English markets. In certain countries, such as Denmark, agricultural cooperatives enabled peasants to transform themselves into modern farmers. The benefits which may accrue to farmers from cooperation are well proved by considerable experience in many countries throughout the world. The cooperative movement has raised farmers from poverty to affluence, enabled them to compete in adverse trade conditions, and sometimes saved them from bankruptcy. In advanced countries, it has been an important force for encouraging investment in the land, and for progress in farming. Yet people may still doubt its role, and wonder whether it can serve the developing countries as effectively as it has the western world. At the turn of the century, despite living in a poorer climate and on generally less fertile soils than in England, the Danes were exporting profitably into British markets when English farming was verging on insolvency.

Though Danish farms were small, they were able to invest capital in the development of their land and to evolve innovative systems of profitable farming, obtaining a good living even from ten to twenty hectares. Their success depended on owner cultivation (i.e. family farming) and on the universal mortgaging of land to raise working capital. The rural population was well educated, in the general sense, and was interested also in education for its own sake as well as for its usefulness, and was prepared to pay for the privilege. The farming itself was *not* cooperative, but credit, purchasing, processing and marketing were. In effect this provided the farmer with vertical integration of the agricultural industry, so that he enjoyed the benefit from profits at whatever level they occurred, via dividends paid to him. Capital could be obtained from cooperative banks, by means of bonds underwritten cooperatively, mortgaging up to about sixty per cent of the value of a property. Cooperatives assisted with the purchasing of feedstuffs and fertilisers, with the hire of machines and equipment, with the breeding of livestock, with the quality control of farm produce, with milling, dairying, slaughtering and processing, and with marketing and exporting the produce. The farmer was saved the time and expense of marketing for himself, and at the same time obtained assured prices and demand.

The policy of the Danish Government had been to foster owner cultivation of the land, by offering loans at a low rate of interest (4%) to individuals with a farming background to buy land, on the security of the land itself. The purchaser was himself responsible for finding a willing seller of land. There was an upper ceiling to each loan, so that farms remained comparatively small. Up to nine-tenths of the value of a farm could be borrowed, the farmer having to provide the remaining tenth for himself as working capital to stock and equip the farm. The debt was repaid over 98 years, but during the initial five years the burden was eased by the farmer having only to pay the interest on the loan, while he established himself, before beginning to liquidate the capital sum borrowed. It was discovered that there would have been less business failure if a farmer had been required to find thirty per cent of the value of the farm, or at least twenty per cent, in the first instance. The excessive proportion loaned by Government also had the undesirable effect of causing an inflation in land values. Regulation of farm size can be assisted by laws of inheritance. On the death of a land owner the property is divided equally among the children, unless the area of the farm is below a certain limit, when it could be left to one child who has to compensate the others. For larger holdings, one child could take over the share of other children, provided he has their agreement and compensates them. The English custom of bequeathing land to only one of the children, generally the eldest son, was not general on the Continent of Europe.

Sweden followed Denmark in the development of farming cooperatives. Earlier, in about 1800, the Swedish government had forced the consolidation of fragmented land ownership, and through acts of legislation had compelled people to leave their villages and move out into open country with their buildings, where each got his own land all in one piece. At first there was considerable opposition from the rural population against such compulsory methods, but this did not persist after the change had proved to be more efficient. In some respects there were drawbacks: traditional cooperation and communal feeling were broken, but the change made better agricultural methods possible, to the great advantage of the farming industry. Over eighty per cent of farmers were owner occupiers. Many improvements were introduced, such as tile drainage, development of mixed farming, introduction of machinery and new implements, and improvement of livestock. The cooperatives also helped to spread newly acquired scientific and technical knowledge to the general farming population. The integration of agricultural industry gave other advantages: for example the tops of sugar beet and waste from crushing could be returned to the farmer to feed his livestock, adding to the profitability of his operations. In more northerly parts of Sweden farms contained woodland as well as arable and pasture, and there were also forestry cooperatives to assist in working them. This was best done during the long winter months, when the soil was frozen and protected by snow, which made it easier to skid out the logs. It also provided employment for the farmer and his draught animals during the period while the weather was unsuitable for field cultivation or for growing crops.

Agricultural cooperation in Ireland

In Ireland, the 1903 Land Purchase Act transformed the Irish tenant into a peasant proprietor. Horace Plunkett organised cooperative societies among Irish farmers, drawing on ideas from Denmark,

Sweden and Germany. The economic activities of the cooperatives were supplemented by education and training of farmers. Abused by history, ignorant and often illiterate, with little agricultural skill, and selling into markets flooded with cheap foodstuffs from overseas, the Irish peasant was now helped to emerge from indigence and semi-starvation to become the modern business farmer of today. Margaret Digby writes, 'The movement which took credit as its point of departure was addressed primarily to those who were penniless, inexperienced and often illiterate. It educated them in the use of money, in businesslike ways, in the understanding of simple economic problems. At the same time, it inculcated a social sense, tolerance, neighbourly solidarity, a desire for education, and the elements of democratic responsibility . . . It was the farmer's answer to the world of big business that surrounded him and his determination himself to break into that world, and be at once producer, manufacturer and salesman.'[17]

Plunkett had some difficulty harmonising his role as head of a government department and promoter of an organisation pledged to voluntary cooperation among farmers. The marketing of Irish agricultural produce was one of the problems confronting the Department (of Agriculture and Technical Instruction). In Plunkett's opinion agricultural cooperation could do more for its solution than the Department. Much that was expected from Government could really only be achieved efficiently and economically by the producers themselves. Plunkett never wavered from what he called 'the cold principle of self-help', and he believed that experience had shown this to be far more important in a poor country than a rich one, because 'everyone helps himself, often to other people's property in a large way'. Plunkett's friendship with Theodore Roosevelt, stemming from their experience as 'cow punchers', was instrumental in the growth of agricultural cooperation in the United States of America. Gifford Pinchot, Commissioner for Forestry, wrote to Plunkett, 'Some of these days it will be known that you [Plunkett] are the man who stirred up the whole movement in America'.[18]

Neglect of cooperation in England

In England, at that time, it was another story. English farmers had been the first in Europe voluntarily to free land from communal

17 Digby, Margaret (1970) *Agricultural cooperation in the Commonwealth,* Blackwell, Oxford (p. 222)
18 Digby, Margaret (1949) *Horace Plunkett,* Blackwell, Oxford (p. 314)

rights, in order to consolidate land ownership and to establish freehold tenure. The result had been an earlier flowering of agricultural investment and development. Subsequently, ownership of a great part of the land had passed out of the hands of the original peasant proprietors into those of large land owners, many of whom had earned their wealth in commerce and industry. English farmers could see the success of the Danes, but seemed congenitally unable to emulate them. Haggard wrote, 'To secure success I imagine that common effort and mutual support would be necessary, and to such things our farmers are not accustomed. Many of them to all appearance would prefer individual failure to the achievement of corporate victory.'[19] There were also other reasons. English farmers were generally not owner farmers, nor were they working farmers, but tended to see themselves as managers of labour. Unless a farmer owns his own land he cannot raise capital for investment from mortgage, and in fact, unlike their Danish counterparts, English farmers did not commonly resort to raising capital by mortgage. Even if the farmer could raise the money, he would not wish to see increased profits swallowed up by higher rents paid to the landlord. Moreover, the economic predominance of Britain as an industrial power, at that time, had led to an overvalued currency which, combined with a national policy of free trade, favoured agricultural importation over domestic production.

Russell wrote, 'The workers in the towns clamoured for cheap food. Let it come from anywhere, they must be fed. The manufacturers backed them up because they knew there were two alternatives to be faced. The purchasing power of money, its capacity to buy bread and meat, had to be increased, or else wages had to be raised and profits eaten into. They preferred to increase the purchasing power of money, and all the business skill and organising power of Great Britain was used to flood the market with corn and beef, with fruit, and farm and dairy produce. It mattered not at all to them that, in the face of this competition, agriculture rotted away outside their own cities, and that the farmers in more distant countries broke up more virgin soil, while at their own doors the land went back to nature . . . It was immediately easier to invite the world to send its supplies of food than it was to develop the natural prodigious food-producing capacity of the fields about them. The farmers were unorganised. They had no trade union, no powerful voice to plead in their behalf, and today [i.e. 1912] the deer forests in Scotland, the game preserves in England, the deserts of

19 Haggard, H. Rider (1899) *A farmer's year,* Longmans Green, London

grass in Ireland, are gigantic illustrations of the desolation and decay which falls on the industry when men work alone and are not united to protect their interest ... Agriculture separates its workers, while the factories and mills bring their workers together. Because of this isolation of the workers in the field, because each man has his plot of earth to till, and because he is made more or less solitary, there is all the more need for organisation ... [Farmers] soon realised that with the union of men to help each other came the promise and potency of a progress inconceivable before; that with more economic business methods, with cheapness in purchase, combination in sale, with science in the farm and dairy, with expensive machinery cooperatively owned, and with the complete control of their own industry – farmers could create and retain a communal wealth which could purchase for them the comforts and some of the luxuries of civilisation.'[20]

Russell also wrote, 'There is the cooperative system, as we find it in Germany, with associations based on mutual liability, where mutual trust is engendered based on long experience of each other's character. ... Wherever there is mutual aid, wherever there is constant give and take, wherever the prosperity of the individual depends directly and obviously on the prosperity of the community about him, there the social order tends to produce fine types of character, with a devotion to public ideas; and this is the real object of all government ... The worst thing which can happen to a social community is to have no social order at all, where every man is for himself and the devil may take the hindermost. Generally in such a community he takes the front rank as well as the stragglers. The phrase, "Every man for himself", is one of the maxims in the gospel according to Beelzebub. The devil's game with men is to divide and conquer them. Isolate your man from obligations to a social order and in most cases his soul drops into the pit like a rotten apple from the Tree of Life ... Let one trader in a street adulterate his goods and sell them cheap and get customers. If unchecked, in five years' time half the traders in his neighbourhood will be adulterating what they sell in order to compete and live ... A social order should provide for three things – for economic development, for political stability, and a desirable social life.' He continues that, in Ireland, 'The [cooperative] societies are controlled by committees elected by the members, and in a decade or so, instead of the dislocation and separation of interests which has been so disastrous in its effects, instead of innumerable petty businesses all

20 Russell, G.W. (1912) *Cooperation and nationality,* Maunsell & Co, Dublin, reissued by the Irish Academic Press in 1982 (p. 103)

striving for their own rather than the general welfare, there will be in each parish one large association able to pay well for expert management, with complete control over all processes of purchase, manufacture and sale, and run by the farmers with the energy of self-interest. These district associations are rapidly linking themselves on to large federations for purchase and sale, which are again controlled by representatives of the societies, and through these the farmers are able to act powerfully in the market.'

The cooperative idea was born in England in the early nineteenth century, stemming from the philosophy of Robert Owen, who wished to incorporate the best of rural traditions into urban industrial life. The Danes had based their cooperatives on the principles of the Rochdale flannel weavers. Briefly these were:

- membership of the society open to any man or woman of good character;
- all members together to share the risk, and to make more or less equal contributions to the resources of the society;
- democratic government of the society, with equal voting rights for all members;
- distribution of dividends – i.e. of surplus earned in proportion to the business done by each member.

The obligation of each member to do all his business with the society was assumed to be morally binding.

Margaret Digby wrote, 'Agricultural cooperation, as it has grown and spread from country to country over the last 100 years has been first and foremost a way of strengthening the economy and developing the resources of the independent farm. Farmers have united in cooperative societies in order to obtain credit and invest savings, to provide themselves with the tools and raw materials of their industry, to process and market the crops and to make available a range of essential services from insurance to electricity. Most countries where agriculture is technically advanced owe much to their progress in this system. It is well-known and well-tested. Anyone describing it can write with confidence of well-known facts.'[21] Dumont wrote, 'The success [in Scandinavia] also depended on cooperative leaders who were honest, competent, dynamic and dedicated to their compatriots; and on their side the members were disciplined and understood that they had to support the cooperative even when it was not to their immediate advantage.'[22]

21 Digby, Margaret (1957) *Cooperatives and land use,* FAO Agricultural Development Paper No. 61 (p. 115)
22 Dumont, Rene (1966) *False start in Africa,* Andre Deutsch

Sir Daniel Hall commented, 'Combinations to sell the products of agriculture had never made much headway among English farmers ... individuals by temperament and tradition ... Such cooperative societies as existed were mainly concerned in the purchase of commodities required by farmers, and various attempts to institute collective selling had revealed the fatal sources of weakness in voluntary cooperation – the difficulty of retaining the loyalty of the partners in the enterprise.'[23] A solution was found in the Agricultural Marketing Acts of 1931 and 1933 which enabled a majority of producers of any agricultural commodity (that is at least two-thirds) to compel the adhesion of all the producers by the promotion of a Marketing Board. Government put pressure on farmers by refusing assistance to agriculture except via the Marketing Boards. Once constituted a Board becomes the selling agency for the product, and no sales are permitted except by registered producers on terms laid down by the Board. The Board itself can buy and sell and conduct any manufacturing or preparation for the market. It is a fundamental policy that the Boards should be controlled by representatives of the farmers. Boards included the Milk Marketing Board, the Egg Marketing Board and the Pig and Bacon Marketing Board.

Recent developments in English farming

During the past 50 years, significant changes have occurred in farm management as a result of the effects of Government fiscal policies on the inheritance and ownership of land. Today, two-thirds of farmers are owner occupiers, and about the same proportion are family farmers, that is working without hired labour. This is a very different situation from that described by Dumont in 1954. About half of the farming population is descended in an unbroken line from the original peasant proprietors, though today less than two per cent of the total working population is employed on the land. Farms are still relatively small, the mode being about 100 hectares. England has lost its industrial pre-eminence, and joined the European Economic Community (EEC), which offers a system of guaranteed prices and markets for farmers, together with grants and tax relief to encourage development of agricultural land. In fact, this is financed by the State out of earnings from commerce and manufacturing industry. Through Farmers' Unions, farmers have become a political force; in part this is

23 Hall, Sir Daniel quoted in Ernle, Lord (6th Ed. 1961) *English farming – past and present,* Heinemann & Cass (p. 559)

a riposte to the Trade Unions of manufacturing industry, whose interests may be antagonistic to those of farmers, for example in the desire for cheap food. In England the functions of the cooperatives have been, to some extent, replaced by the Marketing Boards, and by the development of specialist firms that may be contracted to drain or plough fields, to spray or harvest crops, to trim hedges or to manage woodland, etc. These services are important to the individual owner cultivator, who may employ no permanent hired labour, and whose farm is not large enough to justify purchasing expensive specialised machinery.

The systems of farming developed after land enclosure became general by the mid 19th century. Though traction depended mainly on horses, which had ousted oxen as the main working animals on farms, machines progressively replaced manual labour. In the latter part of the century steam ploughs were introduced and could be hired by farmers. Steam engines were too heavy to be allowed across the fields, and two engines trundled along the opposite sides of a field, hauling a reversible plough to and fro by means of chains stretched between them. During the First World War the tractor began to replace the horse, particularly for ploughing, because of the loss of both men and horses to the war fronts. Horses continued to be used for drilling, harrowing, raking, reaping and binding etc, and in the 1930s horses were as numerous on the farms as they had ever been. The maintenance of soil fertility continued to depend on mixed farming, that is combinations of arable cropping and livestock. Dairy farming was important, being less vulnerable to overseas competition, at that time, due to the perishability of milk, and aided by the Milk Marketing Board.

Since the Second World War there has been another revolution in agriculture. This was a consequence of the general use of mechanical cultivation, cheap artificial fertilisers, and the introduction of chemical pesticides and selective herbicides. These have made the rotation of crops and mixed farming less important. They have promoted specialised farming, the farmer concentrating on a particular crop, or a particular breed of livestock, or even on particular stages of livestock production. The horse has virtually disappeared as a working animal. In 1936, Sir Daniel Hall wrote, "The old mixed farming is being exchanged for businesses chiefly concerned with one or two kinds of production – milk, pigs, fruit, poultry etc. Extreme cases may be noted of milk producers who do not breed, of egg producers who buy all their laying birds. It is expecting much that a man may become expert in all branches of farming, for it is no easy task to accumulate full experience of even one or two. The

Cutting corn with a horse-drawn cutter and binder

Cutting corn with a combine harvester

economic organisation of labour is facilitated when a single object is being pursued. The risk of having all one's ventures in a single shift is being minimised by the Marketing Boards, which have reduced the excessive fluctuations in prices ... the creation of the Marketing Boards marks a new relation between the State and farmers.'[24] It may be commented that another factor has been the rising expense of labour, and its replacement by capital intensive machinery. This reduced the number of people, and hence the variety of skills, employed on each farm.

Since the 1930s, scientific breeding of crops and livestock has made an important contribution to agricultural production. In England, between 1939 and 1979, average yields of wheat, potatoes and milk have more or less doubled. Farming has become a highly capitalised industry, and a farm of 100 hectares may represent an investment of £500 000 (US$700 000) in land, buildings, machinery and livestock. Government has encouraged investment in farming by tax concessions, price guarantees and grants. Businessmen have introduced into farming industrial concepts of management and production, together with commercial methods of accounting and financial control. Even the computer has recently begun to invade the farm. Farmers have had to adopt these techniques, to improve their own efficiency, or go out of business. Fortunately farmers do not indulge in trade secrets! In effect a farmer has become an industrial worker whose occupation keeps him in the countryside. Like the urban worker, he buys his food in the supermarket, and the quality of his life differs only in more varied work, and perhaps in less leisure. Farming has become directed at the efficient production of bulk materials in uniform grade, suitable for industrial processing, and making possible high living standards for ordinary people, the majority of whom are employed in commercial and industrial activity in the towns.

Lessons for agricultural reform in West Africa

In well-populated regions of West Africa, the area under bush fallow is more than three times that cultivated. The productive use of this land could make a tremendous contribution to the economy of farming. Where grassland forms the natural fallow, it is possible that integration of animal husbandry and arable cropping may provide a

24 Hall, Sir Daniel quoted in Ernle, Lord (6th Ed. 1961) *English Farming – past and present,* Heinemann & Frank Cass (p. 559)

solution to permanent farming as it did in England. The ability of tropical soils to hold nutrients – and hence the effectiveness of fertiliser applications – depends chiefly on the humus content of the soil. Also the humus is important for maintaining the crumb structure and stability of the topsoil. Humus depletes rapidly following clearance of natural vegetation. In the humid tropics the fallow may be trees, and the organic matter content of the soil can be difficult to sustain without a tree phase. In West Africa, foresters have long experience of cultivating arable crops to assist establishment of timber crop plantations. Roose shows that other perennial crops, such as bananas, cocoa, oil palm and rubber may also conserve the soil from erosion and loss of fertility, though at a lower level than forest growth.[25] The writer's personal experience in Nigeria, using combinations of arable cropping with establishment of timber crops, and a visit to Jari Agro-Florestal Ltda in Amazonia, where pasture was sown in areas being reforested with pines, making possible cattle raising and paper pulp production at the same time, have convinced him of the importance of integrated farming for profitable land use in the tropics. There is an account from 1925 of a Togolese, Mr Olympio, who many years before had gone to England as a ship's boy and been educated in Bristol, and who had 'a fine and well stocked farm [near Lomé] . . . The farm, planted and cultivated under the German system, was interesting and provided a further example of their efficiency. It was in the main a coconut or copra farm. An essential method for obtaining the best commercial results from these plantations is the maintenance of a large herd of cattle to be tethered nightly and systematically at the foot of the trees through the run of the groves; one animal to each tree. In this way, and thus manuring the palms, a mixed basis of farm is run with arable land sufficient adequately to augment the food for the cattle provided by nature in the rough grass.'[26] It may be noted that, at that time, the tethering of livestock and in particular cows and goats, was widely practised on the Continent of Europe, and made possible a highly efficient use of pasture. The system is, however, very expensive in labour, and has in more recent years been progressively abandoned.

At the International Institute of Tropical Agriculture (IITA), in Ibadan, one sees the experimental development of 'alley cropping'. Hedges of *Leucaena* or *Gliricidia*, which provide fodder and fuel-

25 Roose, E.J. (1982) *Runoff and erosion after clearing, depending on type of crop in Western Africa,* paper at symposium on *Land Clearing and Development in Relation to Environmental Protection in the Humid and Subhumid Tropics,* International Institute of Tropical Agriculture, Ibadan
26 Gray, Frank (1928) *My two African journeys,* Methuen (p. 271)

wood, have four metres wide interrows of maize. Other examples could be cited. Such systems, besides keeping the land in permanent production, add cream to the cake – because the amalgamation of activities is more profitable than where they are conducted alone. Also the farmer becomes less vulnerable to the vicissitudes of a particular crop, and a judicious combination may enable a better spread of the work load on men and machines during the year. At Manaus, in Amazonia, the only profitable farms were those of Japanese descended Brazilians, who combined battery raising of chickens with growing vegetables and peppers, by utilising the chicken manure as fertiliser. The implications for farming are discussed in greater detail in Chapter 6.

Problems of agricultural cooperation in West Africa

Can agricultural cooperation do for West Africa what it has achieved in Western Europe? A problem is that it is seen as a part of a political philosophy, and as an end in itself, rather than as a means to an end. In 1865, John Stuart Mill wrote that the success of cooperation would bring about 'a moral revolution in society, the healing of the standing feud between capital and labour; the transformation of human life from a conflict of classes struggling for opposite interests to a friendly rivalry in the pursuit of a common good to all; the elevation of the dignity of labour and a new sense of security and independence in this labouring class.'[27] However, in 1914, William Ashley wrote, 'But these anticipations have been grievously disappointed. Hundreds of experiments have been made, and there is a noble story to tell of persistence and self-denial in the scraping together of capital, but undertakings of cooperative production in Mill's sense have without exception failed completely, either from the business or from the cooperative point of view . . . most of these failures were due to mismanagement. The undertakings either did not secure competent managers, usually because they were not ready to pay sufficiently high salaries, or else they quarrelled with them. Industrial self-government proved altogether incompetent to organise production and, what is even more important, to secure a market. Success, where it did come, was equally fatal; the small societies of handicraftsmen, such as most of the early groups really were, became exclusive if they were successful and began to employ outside labour; the large societies formed later in the cotton-spinning industry became mere joint-stock

27 Quoted in Ashley, Sir W. (1914, Ed. 1961) *The economic organisation of England,* Longmans

companies, in which, indeed, working men held shares, but in which the employers and employed ceased to be identical bodies. Capitalism has not, as a fact, been seriously modified by cooperation.'[28]

In Nigeria, the cooperative movement began spontaneously with the founding in 1907 of the Agege Planters' Union. This aimed to provide credit facilities to members, to spread knowledge for improving the quality of cocoa, to regulate labour supply, to construct and improve roads and to arrange direct sale of produce abroad. At the first meeting 400 planters were present, and its success stimulated the founding of other producers' cooperatives in various parts of Western Nigeria. In 1934 the Federated Cooperative Cocoa Marketing Union was formed, through the influence of Akinpelu Obisesan who was himself a cocoa producer and cocoa buyer. However, after the outbreak of the Second World War in 1939, its functions were absorbed by the government Marketing Boards, with their monopoly of exports of cocoa, cotton, groundnuts and palm oil. This was, no doubt, a response to the exigencies of wartime, and also perhaps an indirect effect of the success of the marketing boards in Britain. In 1938, J.U. Eku launched the credit movement in the Eastern Provinces, reorganising the traditional 'Esusu clubs' along cooperative lines. Essentially, the members of the clubs each paid monthly contributions, and each member in turn collected the kitty. The member was expected to use a part of the money to pay for food and drink, to entertain the members, on the day appointed for making contributions and paying out. Thrift and loan societies later gave way to thrift and credit societies. These were mainly used to finance trading, and again played a useful part in reviving commerce in the Eastern States of Nigeria after the end of the civil war. In 1944 the Cooperative Supply Association was set up as an import/wholesale organisation to circumvent the (then) dominant position of foreign firms in commerce. In 1952 the Cooperative Bank was founded with a grant of £1 million from the Western Region Marketing Board, which had been able to siphon off large amounts of money because of its virtual monopoly of the cocoa buying market.

In fact, Government began quite early to intervene in the cooperative movement, enacting the Cooperative Societies Ordinance in 1935, and appointing a Registrar of Cooperatives in 1936. Doubtless this was from the best of motives, in that Government recognised the possibility it offered to promote economic development of the country, so to speak from the 'grass roots', and wished to curb abuse

28 Ashley, Sir W. (1914, Ed. 1961) *The economic organisation of England,* Longmans

of position by officers of societies in which most members were illiterate and ignorant of correct business procedures. In 1944 the Cooperative Federation of Nigeria was founded, initially with Government functionaries holding key offices. However these were withdrawn later, because of the implacable opposition of the members, who saw the cooperative movement rather as a buckler against the domination by Colonial Government and expatriate companies.

The establishment of regional self-government in 1954, if anything intensified political pressures on the cooperative movement. Some regarded it as a means to pursue political aims and as a source of political patronage. Government caused investment in financially unsound ventures, and sometimes failed in the timely settlement of debts to societies, causing financial embarrassment. A number of societies defaulted, because there were members who treated loans as gifts, and as a share of Government booty. Personal interests were allowed to outweigh the interests of the societies, which became engulfed in enormous outstanding and overdue debts. Other societies were cooperatives in name only, and the members were virtually wage-earners in the employ of Government. Many societies were too small to be economically viable. Even successful societies might impose additional strains on the communities in which they were established, either by creating new hierarchies, or alternatively by reinforcing the power of existing hierarchies to an intolerable degree. These could be a result of the officials of the co-operative forming a new power base within the community; or by the chiefs or political bosses heading the cooperative, and dominating the economic as well as the political life of the village.[29]

Cooperation has also been promoted, in effect, as a way to try to escape from the intensifying necessity for land tenure reform: the need to change from various forms of communal land tenure to 'freehold', and to consolidate and enlarge farms in order to make mechanised farming economically feasible. This is proposed either by a complete pooling of land, labour and of the proceeds from the resulting harvest, or by holdings being worked individually with some form of cooperative combination of working and with cooperative services for credit, supply and marketing. Such systems may be seen in Nigeria in the Farm Settlement programmes, and in various Group Farms and Community Farms established during the 1960s, and more recently by *Opticoms* etc. It may not be true to say that undertakings of this type

29 Adeyeye, S.O. (1978) *The cooperative movement in Nigeria,* Philipps University Marburg Gottingen (p. 221)

never succeed, but the records of failure are too numerous to ignore. Even where they seemed successful, they might absorb public funds far beyond what their material achievements justified. In Eastern Europe, where such systems of farming have been an article of political faith for more than half a century, their generally dismal performance cannot be treated as encouragement.

The observations of Margaret Digby on the Mexican experiment in cooperative farming may be relevant. She writes that it was 'interesting because it was neither fully spontaneous nor imposed by an all-powerful State discipline. It is concerned with people who are in the main at a low level of technical and general education, and whose initiative and self-confidence had been sapped by many generations of poverty and semi-serfdom. The lesson of the experiment seems to be that, without costly and constant supervision by experts, and the provision of State subsidised services on a considerable scale, the mere introduction of cooperative ownership, or even actual joint cultivation does not, at any rate for many years, result in improvement of methods, increased production or any marked rise in the standard of living . . .' She also writes, 'In trying to assess the value of cooperation or collective farming for practical ends, such as increased productivity or social welfare, it is well to remember that the original driving force was ideology and that the system has psychological as well as technical implications . . . In general, those to whom the cooperative farm appeals are either intellectuals without previous farming experience; the technically minded, to whom machines and all that goes with them are more important than the bare ownership of land; the more dependent type of sharecropper; and labourers who have no experience of managing farms of their own; or tribal peoples who have no tradition of individual ownership . . . Collective farming on any considerable scale is beyond the organising capacity of the ordinary peasant member . . . There is the opposite danger that to place agriculture in the hands of a large body of officials is merely to create a new and costly bureaucracy without that permanent and personal link . . . from which the better type of landlord or farm manager derive their usefulness . . . Experience shows that labourers and tenants, though they may at first accept land pooling, tend to develop the traditional peasant attitude as their experience increases and to demand the division of the land into holdings. It is possible that the attitude of tribal peoples will undergo a similar change . . .'[30]

30 Digby, Margaret (1957) *Cooperatives and land use,* FAO Agricultural Development Paper No 61 (p. 115)

In Western Europe, Farmers' Unions have played much the same role in representing and promoting the sectional interests of farmers as Trade Unions have for industrial workers. To some extent they may be a protection against the contrary interests of organised labour – for example, in their desire for cheap food. However in Africa there has been a tendency to regard spontaneous farmers' movements, such as the *Agbekoya* movement in Western Nigeria, as a threat to the existing political authorities – though it is alleged that political malcontents stirred up and exploited genuine dissatisfactions among farmers. These unions have either been suppressed, or emasculated to serve the purposes of Government. Indeed it may be a problem for cooperative movements in the developing world, that Governments desire to control and manipulate them. The ideological aspect of 'cooperation' gives the movement political overtones, and therefore it may inevitably attract government interference. Agebodeka writes, 'It should have been clear to any careful observer that by incapacitating the genuine cooperatives and setting up party-affiliated ones, Ghana had missed the road to a stable agricultural development. The sense of belonging that had promoted honest and hard work in the old cooperatives was lacking in the new bodies. With the fall of Nkrumah's government in 1966, The National Liberation Council restored the rights and privileges of the original cooperatives to them. But the harm had been done. The interruption and physical damage had destroyed the fine qualities of cooperation attained by 1960 . . . It is true that there are other forces both inside and outside Ghana which have influenced agricultural decline since the 1960s, but had the cooperatives been allowed to develop normally, agriculture would not have sunk to the depth it has now.'[31]

Cooperatives do not of themselves resolve the problems of land tenure, nor furnish economically viable systems of sustained yield farming – the impediments which bedevil agriculture in West Africa. These must be substantially solved before progress can begin. Successful agricultural cooperation is based on a well-educated rural population, on owner cultivation, and on a system of freehold tenure that makes land a freely negotiable asset available for mortgage. Education is necessary, so that there may be innovative farmers, who understand scientifically the management of the soil, and who are capable of administering a modern business. Owner cultivation necessarily restricts the size of farms, but these have to be large

31 Agebodeka, Francis (1984) *A question of cooperation,* West Africa Magazine No. 3469 (pp.326-328)

enough and sufficiently mechanised (though they may use draught animals) so that the farmer can earn an adequate income. Land tenure has to be flexible, so that farm size may adjust to changes in economic circumstances, and so that land can be readily mortgageable in order to give the farmer access to capital, or to funds that may tide him over a financial difficulty. The price of land depends, of course, on both its inherent and speculative values. The inherent value relies on its income earning capacity, which is at present low; thus, land may be appreciably dependent on its speculative value, which may be higher, for its development.

By means of cooperation, farmers can corporately gear up the value of their land in order to acquire ownership, and thereby control, of the activities ancillary to their industry. These include credit, purchasing, processing and marketing. This obtains the benefit of vertical integration of the industry, so that a farmer can enjoy a return from profits at whatever level they occur. The farming itself is *not* cooperative. Successfully innovative farming depends on the strivings of many individuals, with the freedom to do more or less as they choose, and to take the consequences – whether for better or for worse. There is cooperation, though, to the extent that farmers will share improvements in technical knowledge. Cooperatives assist in the dissemination of these discoveries, and of scientific information. They perform functions that are beyond the reach of most individuals, such as breeding livestock or crops, and carrying out applied research. The farmer entrusts the management of the ancillary cooperatives to specialists in each particular activity. Like shareholders in a joint-stock company, the farmers use their voting power to ensure that the cooperative is managed in their common interest. As in all business enterprise, the executives must be honest and dedicated to their duty, if it is to be carried on successfully. On their part, the cooperative owners (the farmers) must ensure this is done, and in turn have the responsibility, common to all employers, of treating their employees with justice and equity. They should be remunerated adequately, and good performance rewarded with advancement and material benefits. Cooperation gives the petty capitalist protection against big business and, it may be added, against Government. It is not some special dispensation from the laws of economics; success does not depend on starry-eyed philanthropy but on the real advantages of joint venture.

Russell wrote, 'The greatest danger the new rural organisations have to face is interference from the State ..... The State would endeavour to interfere with everything, dominate everything, and break up every organisation of the people which threatened effective opposition to, or control over, the actions of the Mandarins.' He continued, 'The effect of the policy of our present public men is to turn the Irish into a race of economic babies, with their lips for ever nuzzling at the nipples of the State.' Also, 'The State, with its gigantic machinery, its innumerable array of officials, and its power to draw on the public purse . . . is a real danger . . . I am now convinced that there are enough "Toughs" in Ireland to carry through the scheme of rural reorganisation, and to keep the Mandarins and politicians in their proper place: who will see that officials are the servants of the public and not the masters. The Eden of the bureaucrat is the hell of the governed . . . The official classes will, I believe, be much happier serving the public than in setting snares, or inventing schemes, to control industries and movements they had no part in creating, where their interference would be fatal to any fine idealism or noble humanity. The country has seen lately how a great national institution can be degraded in popular estimation by its headship being handed over to an incompetent economist and bitter partisan.'[32]

Other obstacles to agricultural development in Africa

In Africa there may be considerable obstacles to the development of successful agricultural cooperation. The situation with regard to land tenure may be unsatisfactory and confused. For example in Nigeria the 1978 Land Use Decree appears to restrain rather than encourage the development of freehold tenure. Other African countries have also erected legal barriers that effectively prevent the evolution of land tenure. It does appear that agricultural development depends on owner farming, and on land becoming a freely negotiable asset. Government should legislate both to create the framework within which this may occur, and actively promote the change with a view to arriving eventually at a nationwide system of land tenure. This also depends on modes of inheritance, and on limitations on the degrees of amassing or fragmenting of estates, and on the ability of a man to give or transfer land or capital during his own lifetime. Otherwise people should be allowed to change voluntarily at their own pace, subject to the carrot and stick of fiscal measures. For example the imposition of an annual land tax (and/or rates), which weighs most heavily on larger estates, whether owned individually or jointly, would encourage communities or families to subdivide their lands. This would allow piecemeal change, reducing the degree of trauma to the country as a

32 Russell, G.W. (1912) *Cooperation and nationality,* Maunsell & Co, Dublin (reissued in 1982 by the Irish Academic Press) (p. 103)

whole, and giving people time to learn from each other's experiences. We should restore to land law the flexibility, and the ability to evolve to suit changing circumstances, that customary land law enjoyed traditionally.

Mary Kingsley wrote in 1898, concerning the hut tax riots in Sierra Leone, 'One of the root principles of African Law is that the thing you pay for is a thing that is not your own – it is a thing belonging to the person to whom you pay the fee – therefore if you have to pay the government a regular and recurring payment for your hut, it is not your hut, it is the property of the government; and the fact that the government has neither taken this hut from you in war, bought it of you, nor had it given as a gift by you, the owner, vexes you "too much", and makes you, if you are any sort of a man, get a gun. The African understands and accepts taxes on trade, but taxing a man's individual possession is a violation of his idea of property.' The proponents of the tax considered that it would be a benefit to the Africans, in that it would help finance the development of their country. But Mary Kingsley also wrote that we should not engineer our good intentions 'in such a manner as to make them appear tyrannies and hateful to those whom we wish to benefit by them.'[33]

However, the imposition of land tax should perhaps be seen in a different context. It is, in effect, a sort of compensation paid to the community by the individual land owner for the rights that he has appropriated from it; and particularly to offset the expense that now devolves on the community of providing for its indigent members. Formerly the latter would have been given a piece of land as their means of livelihood; but this is no longer feasible as the population grows too large, and the development value of land becomes too great. When this happens, the community must seek to guarantee the livelihood of its members in other ways, and these necessitate spending money. It is right that larger land owners should pay proportionately more, because they displace larger numbers of people.

A fear is that educating rural populations merely accelerates the drift of young people to the towns. To some extent this is inevitable. The Universal Primary Education programme, begun throughout Nigeria in 1976, will establish an increasingly literate rural population, at least in the vernacular languages. In the south of the country, more than half the population is already literate or semi-literate – though to

33 Kingsley, Mary H. (1898) letter to the *Spectator,* March 1898, quoted in Gwynn, S. (1933) *The life of Mary Kingsley*, Macmillan (p. 285)

be literate is not necessarily to be educated. Progressive farming requires educated farmers. However, the drift may be partly countered by providing regular wage employment in rural areas, for example by a modicum of plantation agriculture and of integrated agricultural or forest industry. This can help to improve the rural infrastructure, by subsidising the costs of providing education, health services, electricity, water etc, in rural areas. At the same time, though, legislation and active intervention on the part of Government may be needed to ensure that rural wage-earning labour is not exploited.

Peasant farming *will* become replaced by business farming, but this metamorphosis depends on the evolution of profitable systems of sustained yield farming, that will improve the productive capacity of the soil and keep it in permanent use. In the humid lowland tropics there is little experience with such methods of farming at present. These will either dispense with the fallow, or use it productively; but they have to overcome the greater difficulties of soil management in the lowland tropics. Men who have accumulated capital in commerce and industry should be encouraged to invest in farming. The recent depression in commerce, and the increasing insecurity of industry based on imported raw materials, is likely to stimulate their interest in agriculture. They are men of ideas, but often still sufficiently close to their peasant roots. They will probably lose money at first, and will be unwilling to comply unless they can recoup their investment via the enhanced land values that will ensue from the development of profitable farming systems, and from improvement in soil productivity.

The solution to current difficulties in the agriculture of lowland tropical Africa is not beyond the wit of men. Admittedly the biological constraints are much harder than those faced by Europe at the same stage in development, because soils in the humid tropics are generally less stable and less retentive of nutrients and water than those in temperate countries. Moreover, Africa can no longer work out its problems in relative isolation as England was able to do; but scientific and economic understanding are already sufficient for progress to begin. Oyenuga writes, 'Agriculture in the countries of western Europe did not progress substantially from the level of subsistence economy until the full impact of the industrial revolution and its resultant technological influences were felt. It is of interest to note that crop yields and to a lesser extent animal productivity were little different in pre-industrial Europe from what they are now in Nigeria and other countries of West Africa. The yield of grains then varied between 800 and 1500 lb per acre compared with the present range of 2000 and 3000 lb. [That is 900 to 1700 kg/ha versus 2250 to

3350 kg/ha in 1967] ... The average size of farms and the total area cultivated in each country were increased; production per unit area of farmland was raised and a substantial rise in total productivity resulted.'[34] Given the possibility of sufficiently remunerative farming, then the food problem could solve itself.

Conclusions

Nowadays 'Agrarian revolution' may be regarded as an old fashioned term, but it means exactly what it says. Peasant systems of shifting cultivation are insufficiently remunerative and excessively wasteful in their use of land. Consequently they do not produce enough food for expanding populations; at the same time peasant society no longer satisfies social needs; major reform becomes inescapable. Nigeria is the country most likely to lead West Africa through its agrarian revolution, just as England led Europe. The reasons are the size and density of population in Nigeria, the degree of urbanisation, and the wealth of its economy. As in England, the requirement from agricultural reform is *not* to obtain a more equitable distribution of land among an essentially agricultural population – which could achieve little – but to boost agricultural productivity to meet the needs of an increasingly urban society. For the peasants of England the reforms were a tremendous leap in the dark, and it took great courage to yield up the known for the unknown; but they released the flood of ideas and initiative that were necessary to solve the problems, and underpinned Britain's industrial growth. There was a stark choice between gradual impoverishment under an unreformed system, or a change which might appear to have gambled with the future, but which offered a prospect of revitalisation. The need was to sweep away traditional restraints that had guaranteed subsistence to peasant communities but that could no longer feed a nation. In West Africa the difficulties may seem even more intractable than they were in England at that time, but the oil wealth now available to Nigeria may help to finance the change, and may even help to avoid the desperate hardships suffered by so many people in England as a result of land enclosure.

What concepts are revealed by the earlier experience which may be of use in Africa?

1 The problems of land tenure must be resolved in order to disencumber land of communal rights.

34 Oyenuga, V.A. (1967) *Agriculture in Nigeria,* FAO, Rome (p. 308)

2 Methods of permanent agriculture must be understood before progress can begin.
3 Farming has to become adequately profitable, and the returns on capital invested at least comparable with those obtainable in commerce and industry generally.
4 There needs to be an audience of farmers who are receptive to modern ideas, and to experimental findings, and who are capable of putting them into effect.

Land cannot be treated indefinitely as a 'dole' to furnish subsistence to those not otherwise gainfully employed. It should be realised that, in Nigeria for example, *de facto* alienation of the majority of people from the land is occurring already. Human endeavour will solve the problems, unless it is prevented from doing so. The task may be daunting, but the alternative is to risk slipping into a slough from which it may be difficult ever to emerge.

6 The implications for farming

The problem of traditional farming is that its flexibility is too limited; but there is no virtue in replacing it with systems that will themselves prove restrictive, and unable to evolve. For example, farm size must be able to adjust to the optimum for different cropping systems, and to changes in market forces. Certainly the biological constraints are harder to master than those which confronted the farmer in Europe; but knowledge and technical capability are already sufficient.

'The land can rarely return three clear living profits – one to the owner, one to the farmer, and one to the labourer. The owner should be the farmer, or the farmer the owner.'[1] At least in Nigeria, it is doubtful if, within a generation, there will be commercial farmers who are also working farmers. The successful self-improved peasant would rather regard himself as a manager of labour, as would also the farmer who is a trained agriculturist. Former peasant farmers, illiterate or part literate, and often aged in their fifties or sixties, are farming larger areas using traditional or semi-traditional farming techniques, and relying mainly on hired labour for the farming operations. They may use fertilisers and/or tractors to plough their land. The labour often comes from neighbouring countries in a less fortunate economic situation; or it may be migratory labour, for example from the far north when the long dry season prevents farm work, or local youths who want to earn some pocket money. The peasant adopts the style of a petty contractor or Public Works foreman – men who have for many years carried out minor building operations and road works. However, despite the larger area farmed, his earnings are still not much more than those of a Government labourer, perhaps between US$1000 and $5000 per year (in Nigeria); and in purchasing power not a great deal above what he earned when he was cultivating a smaller area of land by his own efforts. The reason is that the work is still mainly manual, and he now has the added cost of hired labour. Time

1 Haggard, H. Rider (1899) *A farmer's year*, Longmans Green, London

will correct this situation, as it did in England, because the advantages of owner cultivation are too strong to be ignored – provided that inheritance and fiscal treatment prevent land falling into too few hands. However, the 'manager-farmer' may persist until labour becomes so expensive and scarce that the only way to get a job done will be to do it oneself. This will also depend on increased mechanisation of farming, on industry soaking up the surplus labour, and on restricting population growth. Meanwhile farming must be protected from the efficient owner cultivator overseas, if expenditure on food imports is to be controlled. Success will hinge on the treatment of farmers for tax purposes, and on the imposition of customs duties on imported foodstuffs – and the will to see that they are actually paid. Import licences may merely place a goldmine both in the hands of those who are fortunate in being responsible for issuing the licences, and in those of the lucky individuals who receive them.

The development of farming depends on farmers having access to credit, and on those who can use credit properly having access to land. If land is in the wrong hands it will not develop. If it is in the right hands, but they have no money to invest, then it still will not develop. The obvious conclusion is that land has to be readily transferable between individuals, so that it may pass into the possession of those who can use it best. If land can be mortgaged, then a means to obtain capital is readily open to those that have land but no money. However this depends on land being a negotiable asset. This in turn depends firstly on the conveyancing of land being quick, easy and, preferably, cheap, and secondly on there being a market for land. The latter depends to some extent on the existence of speculation in land. It is perfectly clear that 'character lending' is unsatisfactory, and perhaps not enough people can prove adequately impeccable characters to allow sufficient investment in agriculture by this means. If people have a good title to land, and the banks have a good security, then the character of the owners may be a secondary consideration. Whether due to bad character, bad judgement or bad luck, if a man loses his land, or a part of it, that he has laid down as security, then the same land may pass to someone who has good character, good judgement or good luck – or else people will learn these virtues! Tough though this is, nevertheless it gives more chance of better farming and of the nation learning to feed itself. The social problems that may ensue from such a policy have to be faced and resolved by means which will allow land to develop and, in effect, to finance its own development. The only alternative, that I am aware of, is for land to be nationalised and for the State to farm it. Even if this were politically feasible, previous experience in Africa does not encourage an optimistic view of State farming for economic production – indeed it appears to be a shortcut to disaster! Finding out how to farm will cost money, and the entrepreneur will want to be able to recover the capital invested in solving the problems of new systems of farming. His investment is speculative, and his only prospect of retrieving his money may be from the enhanced land values that will result from development of profitable agriculture – another reason why land must be freely negotiable.

The need for innovation

Private enterprise, by its very nature, is enterprising and innovative; we need innovative farmers, so that many minds may have the liberty and motivation to solve the problems. Innovation depends on education, and this means education in rural areas, but people, if they are educated, will abandon farming unless they can earn sufficient from it to justify the expense of their own training. Nor would it be worth educating them, unless the cost of their education could be recouped via greater productivity, which in itself leads to greater earnings. Thus progress in agriculture must come from fewer farmers feeding more people. There is also need to attract 'new men' from commerce and industry, because they have already grasped the 'business' concept, which will turn farming into commercial enterprise. The entry of these people into farming will also go some way to restoring the respectability of farming, and to improve its status, so that it is not simply the last resort of the unlucky, the unenterprising and the uneducated. Perhaps the term *peasant* (or *bushman* in West African parlance) may lose its abusive undertones? If money is to be attracted from commerce and industry into farming, then agriculture must be at least comparably profitable with commerce and industry. Land has an advantage, in that it has an inherent value, and can be used as a business asset in order to borrow money and to gear up the value of a business. This helps to make land more attractive as an investment – for the same reasons that businessmen use their surplus earnings to build houses or business premises. Moreover, in this form, it is less vulnerable than money in the bank to importuning kinsfolk.

The wealthy businessman can make a particularly useful contribution to the development of farming, because he can spend his money, more or less, as he likes. The land is its own security, and he will have less difficulty convincing the bank manager of the wisdom of his venture. The problems of farming are going to be solved by the seat of the farmer's pants (that is by moment to moment decisions, as in racing a horse or a car) and success will come by his backing his hunches with money – rather than by the sorts of detailed projects that would convince an accountant, or some technical aid administrator in a distant capital – and which may fail for all that. Nature, inconsistent and inconsiderate, is the final arbiter. Also, large joint stock companies are not often suited to situations requiring a high degree of innovation, judgement and luck. Business executives are answerable with their jobs to the shareholders, and they are paid to take only calculable risks. They usually succeed by the professionalism of their management, and by using well-proved techniques more efficiently than anyone else. Government technical aid is quite capable of throwing money away, but a discussion of its inherent defects will be left until the final chapter.

The small farmer cannot afford large risks, but he can be innovative in small progressive ways. Africans, in common with most peasant peoples, are generally unaccustomed to what one might call professional innovation. They can be difficult to change in their ways, reverting to time-honoured methods as soon as supervision is removed. With sufficient training, though, they can be taught to understand innovation, and to accept it as a way of life, as I have found from personal experience. People can become familiar with a situation of perpetual change. Nevertheless this is a particular problem for managers in Africa, and it can often be easier to introduce completely new methods than to try to modify old ways.

The need for agricultural cooperation

Cooperatives can offer the small farmer some of the advantages, enjoyed by big businessmen, that he otherwise lacks. By means of cooperatives he can acquire specialised professional management for activities ancillary to his operations, and he can share in the profits from agricultural industry at whatever level they occur. Cooperatives can help initially with purchasing and marketing, and then with slaughtering and processing. They can also help to supply other services, such as electricity and insurance, breeding of livestock and crops, laboratory facilities and applied research, education and training. The performance of cooperatives may be monitored by Government, and encouraged by subventions and grants. However, the general experience makes it crystal clear that successful agricultural cooperation depends on the independent owner-farmer, and on having educated farmers with a strong spirit of common purpose and honest dealing. If farmers are illiterate and ignorant, then they risk either being cheated by the few individuals capable of managing the cooperative or, if Government virtually runs it for them, of seeing it as an organ of government – and notoriously what belongs to everybody is the responsibility of nobody. The ordinary man has always had a cavalier attitude to the property of government – illustrated by such expressions as to *liberate* or to *win*, meaning to purloin government property. Indeed, in Africa, it has not always been the ordinary man who has treated government belongings in this way. Farmers must control the operations of the cooperatives themselves and in their own interests, by their voting power. It is also apparent that the ability to mortgage land has been an important factor in providing the financial base for agricultural cooperation.

In England, marketing boards were set up in order to force a cooperative structure on the unwilling farmers. A marketing board is a sort of compulsory cooperation, so that a recalcitrant minority can be compelled by legislation to conform with the majority – but otherwise the same tenets apply that affect voluntary cooperation. However, in Africa, the marketing boards have usurped the functions of cooperatives in order to enable Government to try to control an important source of national wealth, and thereby to siphon off part of it into Government coffers, or sometimes, if possible, into political party coffers. The consequences have been disastrous both for agriculture and for the nation. There is no substitute for spontaneous voluntary cooperation, and certainly not in respect of the moral uplift that it can give to people, as compared with the coercive cooperation of marketing boards. Marketing boards are a second best, originally intended to make obstinate English farmers have some chance of competing with foreign (i.e. cooperative) agriculture in a situation of free trade! Cooperation may be more likely to succeed in Africa, where its roots are based in the descent group or local community, where there is a tradition of communal loyalty. The hierarchies of cooperation can rise towards the apex institutions through local community, district, region, and ultimately nation. *Tribe* may have become a dirty word, but it has been the basis of ancient cooperation, and can be turned to benefit as well as to harm. Ultimately *tribe* will disappear, but *cooperation* will remain.

The need for soil management

Farming is essentially soil management. Soils in the humid lowland tropics are generally more fragile than those of temperate countries.

The critical problems are to check leaching and erosion, and to maintain organic matter content of the soil. Peasant farming was conservationist, and sustained the structure and fertility of the soil, albeit at a lower level than that of virgin soils. The peasant farming systems were regulated by tradition and by a web of communal rights and practices. Modern farming, on the other hand, is a conscious and intelligent operation on the part of an individual farmer to improve soil fertility and productivity, with a view to making farm work more profitable. The farmer needs both the scientific and practical knowledge to enable him to manage the soil to best advantage. For this he also needs the support of laboratories and scientists, and of banking and accountancy, in order to enable him to monitor more rapidly and precisely (a) how well he is succeeding with improving his main resource, the soil, and the quality of his product, and (b) the efficiency with which these are achieved. He has to boost the biological productivity of the soil, so that more of the produce may be drawn off for human consumption. An obvious first step is to discover how either to use the fallow productively, or to dispense with it. In the forest zone the natural fallow is trees, and in the savanna zone it is wooded grassland. The traditional practice of mixed cropping is too complex and difficult to manage economically on an extensive scale, except in very simplified and systematised forms. Originally, foresters tried to simulate the complexity of natural tropical forest in their timber crop plantations, but with almost total lack of success. Despite their misgivings they had eventually to resort to monocultures or to regular mixtures of only two species.

The role of animal husbandry

The crux of the problem is that neither forestry nor staple food crops give sufficient financial return on capital invested, despite the fact that our civilisation cannot exist without them. Staple foods enable us to fill our bellies, and wood pulp (via paper manufacture and printing) to fill our minds. The value of timber still depends largely on the extraction cost from God-given forests, rather than on the expense of producing wood from man-made plantations. The prices of staple foodstuffs must compete with large-scale efficient mechanised production in North America and Australia. Even if home production is protected, the higher prices of domestic foodstuffs will be reflected in higher prices of home manufactured goods, which must also compete with foreign manufacture, and results in other ramifying effects and distortions through the economy.

Mellanby is correct in pointing to livestock as the most potentially profitable form of husbandry. Nigerian businessmen and women have not been slow to take advantage of this possibility. It is to cattle and pigs and poultry that they have turned for agricultural investment. A hectare of cereals may give an annual income of about US$100, and a hectare of tree crop perhaps the same, whereas a hectare of cows or sheep is worth between $200 and $500. However, the cattle are mainly maintained on unimproved pasture, and the food for the pigs and poultry is bought in, and not infrequently imported. When there is a shortage of grain or fishmeal, due to inadequate domestic production, or inability to pay for importation, then the pigs and poultry have to be slaughtered and sold at a loss, or die from starvation. Alternatively, either to feed the populace or to oblige friends, the Government issues special import licences for pork or poultry, and the home producer has animals or birds, ready for the market, that he cannot sell, or can only sell at a considerable loss. These are not theoretical considerations.

The conclusion is obvious. The business farmer must raise the fodder to feed his own animals – and, willy nilly, we are at once into mixed farming. If we are producing fodder crops and livestock at the same time, then we must enclose the land – that is fence or hedge it – if only to control the access of the animals to the fodder. If we fence the land, it is cheaper to plant and maintain hedges than to import creosote and barbed wire, and to obtain posts from a terrain off which the trees may have been cleared already. Electric fences are fine, so long as the electricity is on, and the generator is functioning, and the batteries can be recharged. In Europe, before the Second World War, only a minority of farmers had electricity or tractors, but they still farmed! If there are hedges, one may as well make them useful as well, by incorporating timber trees or fruit trees to provide estate timber and firewood and shelter. The quickthorn, or hawthorn, of England was an ideal hedgeplant. It could be readily struck from cuttings; it was thorny, and relatively quick growing and hardy; it was not poisonous, and it formed a thick hedge that could be coppiced or layered. In Africa, hedge plants with all these properties do not spring immediately to mind. Plants used traditionally as fences include spiny Euphorbias (e.g. *Euphorbia kamerunica* and *E. deightonii*), *Commiphora africana, C. kerstingii* (Myrrh), *Lawsonia inermis* (Henna) and *Moringa oleifera* in savanna areas; and *Newbouldia laevis* and *Dracaena arborea, D. smithii* and *D. perrottetii* in forest areas. In East Africa the cactus *Opuntia vulgaris* (Prickly pear) is used. All can be established from cuttings. Nomadic herdsmen use the entangled thorny branches of *Acacia* species to corral their cattle.

An English hedgerow

Typical Nigerian countryside

Clearly, research is urgently needed to identify potential hedge plants and to try them out as cattle fences. Moreover, if the land is enclosed, and laws of trespass are enforced, then wildlife necessarily receives protection as well, and the farmer may also exploit this. In any event he will certainly have to cull the numbers of wild animals in order to protect his livestock and crops.

More livestock means more dung and more urine to manure the fields, to help sustain both the fertility and humus content of the soil. The numbers of stock that the land may carry will be augmented by pastures of improved grasses and legumes which, besides giving better feed, help to improve the soil. Temporary pastures, or leys, can be used as part of crop rotations to help restore the soil and to clean it of weeds; but if there are more livestock, then there is the problem of feeding and watering them during the dry season. This means either irrigated pasture, or reserving food and stall feeding them during this period of the year. Grain, root crops, hay, straw and silage can be stored to provide fodder. It may be possible to use chopped sugar cane, or waste from a sugar mill, to help make silage. Indeed it is the integration of available waste materials into farming which really adds to the profits, because waste materials can often be had for about the cost of transporting them. At present, in Nigeria, the waste materials from brewing are simply dumped; and West Africans are no mean drinkers of beer! Yet in Europe, these were formerly used to stall feed dairy cattle in the towns, in order to provide milk for urban dwellers. If crops are being grown to feed animals, then the surplus can be sold to feed human beings. This surplus, from an enhanced production, may be more than the yield from an unimproved natural soil. Thus the whole biological productivity of the land is swelled, and what is more, profitably so. At the same time the farmer retains much of the self-reliance and independence which has stood the peasant in such good stead.

Oyenuga writes, 'The value of livestock – cattle, goats, sheep, pigs, poultry and their by-products – and of the activities associated with their productivity, therefore is the smallest of the different sections of the agricultural sector in the Nigerian economy. This shows that until now considerably less attention has been paid to animal husbandry activities than to the production of crops for home consumption and export. This is a rather anomalous situation, typical of underdevelopment, because the value of livestock and livestock products is usually greater in the agricultural economies of the more developed countries . . . This lack of integration of animals with crop husbandry systems has had a deleterious effect on livestock development in Nigeria and is detrimental, in many ways, to agricultural improvement. It prevents animals from contributing to the maintenance and improvement of the fertility of the land...' He also writes, 'The dung from the different classes of livestock provides organic matter valuable for maintaining humic [organic matter] content of the soil, increasing the soil bacteriological population and raising the nitrogen, phosphate and potassium levels. The natural cycle of nature is broken when animals are not used to return to the soil, in a systematic manner, part of the organic material removed by them in their nutrition.' He continues, 'Another factor which is responsible for the low level of animal production has been the small demand for animal products such as meat, milk and eggs, owing to the purchasing power of the Nigerian people... A pound of meat costs nearly twelve times as much as the same weight of Guinea corn or millet. Consequently, those who cannot afford meat tend to purchase and consume only grains or root crops. For these reasons the consumption of meat and fish, of the order of 18 kg/caput/year is very low, when compared with the intakes of certain other countries in the world... Modern animal production requires many kinds of capitalisation. Animals need well-managed pastures, investment in feed purchases, hygienic buildings for the housing and rearing of calves, milking of cows and processing of milk. These have no place in a subsistence economy...'[2]

Tree crops and animal husbandry

Animal husbandry may be combined with tree crops, as well as with arable crops. In Amazonia I saw cattle raised on a pasture of the grasses *Brachiaria humidicola* or *Panicum maximum,* deliberately sown among pines planted for pulpwood, which would take six years to close canopy. It is also possible to run cattle or sheep on cover crops of *Calopogonium*, *Centrosema, Pueraria* or *Stizolobium*, sown under oil palm, or coconuts or rubber.[3] Goats are best stall fed, because of their susceptibility to disease (particularly pleuro-pneumonia); fodder may be produced from hedges of *Gliricidia* or *Leucaena*. The *Gliricidia* readily roots from sticks pushed into the ground, or even from sticks left lying on the ground, and it may be difficult to kill and

2 Oyenuga, V.A. (1967) *Agriculture in Nigeria,* FAO, Rome (p. 308)
3 One hectare of oil palm plantation produces as much oil as 2 hectares of coconut palm, 5 hectares of groundnuts and 10 hectares of soya. The oil from the palms is appreciably more valuable than that from the legumes, and the palms will yield consistently for 20 or more years. However coconuts, groundnuts and soya beans are also valuable sources of protein.

Alley cropping

dry them for firewood. Goats will achieve this by gnawing off the bark, as well as eating the leaves. *Leucaena* foliage contains an alkaloid that causes livestock to lose their hair, if it constitutes more than half their diet. Other leguminous species such as *Albizia, Baphia* and *Pterocarpus,* may also be grown for fodder. A problem may be that fodder trees are deciduous, or flush leaves at intervals, so that there are seasonal fluctuations in foliage production. It may, therefore be necessary to overproduce foliage at some times of year, and to use several species of tree that flush at different periods. If hedge production is combined with alley cropping of arable crops, such as maize or cassava, the surplus growth can be used to mulch the alleys, which are about four metres wide. *Leucaena leucocephala* (a root nodulated legume) was shown to contribute to the soil about 150 kg of nitrogen per hectare yearly, in experimental alley cropping at the International Institute of Tropical Agriculture near Ibadan. Moreover, mulching improves weed control and reduces the need for herbicides; it protects the soil from the impact and erosive effects of rainfall, and prevents the sun from drying out the topsoil; the addition of organic matter encourages earthworm activity and improves the physical condition of the soil. The alleys must be wide enough for machinery to pass between the hedges. However the system may only be suitable where the rainfall is ample, and the water consumption of the hedges does not detract from the growth of the arable crop. Where livestock are stall fed, it is important that their dung and urine are returned to the fields. It is thought that integration of livestock with arable and/or tree crops could be sufficiently profitable for farms of between 20 and 100 hectares in extent.

Fruit trees, such as Mango (*Mangifera indica*), may also be planted round the fields, and wherever shade must be provided for cattle. The Cashew (*Anacardium occidentale*) can be widely grown in both the forest and savanna zones; the kernel of the nut is edible and the shell yields a valuable industrial oil; the swollen fruit stalk is also edible and fruit-like, and can be used to make a refreshing drink. The cortex and kernels of the nuts of the shade tree, Indian almond (*Terminalia catappa*) are also edible. Guavas (*Psidium guajava*) will grow along water courses, and may help to stabilise the banks of irrigation channels. Though these various trees occur widely in West Africa, they are not indigenous. Poplars have been little tried, but may also grow on watersides and on seasonally waterlogged ground; they are quick growing and provide wood that is valuable for the manufacture of match sticks and match boxes. *Acacia albida* is a natural tree of the arid zone, and the fruit pods provide valuable fodder. It is deep rooting, and has the unusual habit of being leafy in the dry season and losing its leaves during the rains. Thus it provides shade for cattle, and shelters the soil during the dry season, but during the rains interplanted arable crops can receive full sun, and the soil is mulched and fertilised by the fallen leaves.[4] In the drier savannas, a useful evergreen shade tree is the Neem (*Azadirachta indica*), a commonly planted exotic. The infusion from leaves or bark is used as a febrifuge; the fruits contain an insecticide that can be used to protect granaries; and the meal obtained by grinding the fruits may be applied as a dressing to improve soil nitrification – enough fruits to treat a hectare may be obtained from a single tree. Indigenous fruit trees which may be suitable for planting include *Chrysophyllum albidum* and *Spondias mombin*, and in drier areas *Sclerocarya birrea* and *Parkia biglobosa*.

It has long been the practice to raise arable crops when establishing tree crop plantations, including timber crops. The arable crops benefit from the residual fertility remaining in the soil when the original vegetation cover is removed, and which would otherwise be uselessly

4 R.W. Fishwick has stated that 15 *Acacia albida* per acre (that is about 40 trees per hectare) in farmland are sufficient to maintain soil fertility.

leached away. The tree crops benefit from the cultivation given to the arable crops, and from the control of weed growth. The returns from both crops help to defray the considerable expense of clearing the land. Where the planted tree crop will regenerate itself by coppicing, it may not make economic sense to uproot the stumps in order to resow arable crops at some later date, because of the cost of replanting them. It may be better simply to allow the trees to resprout from the old rootstocks. The lop and top from exploitation of the tree crop falls on the soil surface, increasing the organic matter content of the topsoil and encouraging earthworm activity. For combinations of food crops and forestry crops to be sufficiently remunerative, much larger areas may have to be farmed than where livestock are incorporated into the system, because of the lower returns per unit area, and may need to be at least 100 hectares.[5]

The role of mechanisation

Dumont wrote, 'Mechanical power is the chief instrument of modern agriculture, and its use must correspond with enlarged fields and bigger farms.'[6] The tendency in Africa has been either to use machinery to extend peasant farming systems, or to try to fit farming systems to the existing machines. This may succeed in the high altitude tropics, where soil conditions approximate to those in temperate countries, but in the lowland tropics – though to a less extent in arid areas – the soils are physically fragile and prone to leaching and erosion.

Topsoils are typically shallow and, particularly in the high forest zone, the experience is that heavy machinery may scrape away the topsoil and compact the subsoil, which impedes drainage and results in waterlogging and reduced yields. Subsoils may become indurated when exposed. Consequently, at least for tree crop plantations, the tendency has been to abandon heavy tractors for land clearing, in

5 Estimates of Financial Rates of Return (FRR) for combinations of maize and cassava with establishment of timber crop plantations of Gmelina or Teak, by J.B. Ball and L.I. Umeh at Ibadan, gave:

planted 1975	Gmelina 5,8%	Teak 4.0%
planted 1980	Gmelina 18.0%	Teak 6.6%

Owing to the continuing rise in relative prices of wood and foodstuffs, the FRR will have also improved.

6 Dumont, René (1957) *Types of rural economy,* Methuen, translated from (1954) *Economic agricole dans le monde,* Paris

favour of chain saws and manual labour. For arable crops, the remaining stumps and debris are a serious obstacle to mechanised operations. Although the use of a crawler tractor fitted with a shear blade, for land clearing, disturbs the soil less and leaves most of the roots *in situ,* nevertheless the effects of fire and the ash may be much less evenly distributed over the terrain than after manual felling. The unburned debris is pushed into windrows, and if possible reburned. The removal of the stumps of larger trees may be difficult after the tree has been cut down, when the trunk cannot be used to lever the stump out of the ground. It may be possible to sow maize, or plant cassava cuttings, directly into ground prepared in this way. If it is necessary to ridge or contour plough the land (e.g. to plant a root crop) the use of a root rake may be essential to remove roots. This disturbs the soil much more severely, and can lead to serious gulleying and sheet erosion. At the International Institute of Tropical Agriculture in Nigeria, terracing and contour bunds were found to be unsuccessful as a soil conservation measure during cultivation. Savanna soils, derived from basement-complex rocks, tend to be compacted, and subsoiling and harrowing may be necessary to promote adequate infiltration of rain – but likewise may result in destabilisation of the soil under arable cropping. It has to be admitted that the existing mechanical methods used expose the soil to severe risks.

The acquisition of the heavy machinery needed for such operations is beyond the means of small-scale farmers. Though, in theory, farming cooperatives or hire units might supply this deficiency, neither has proved satisfactory in practice. Perhaps it is a case of trying to run before one can walk, because the technical and managerial sophistication needed to keep this sort of equipment running is not yet sufficiently widespread. There is much talk of intermediate technology, but intermediate for what? Designers of machinery for small farmers tend to start from the needs of a proposed farming system, as it may have been conceived in a research institution, rather than from whatever best fulfils the needs of the farmer. They might be amazed to see the use to which Nigerian log-takers put a 5-ton Bedford chassis, on the back of which is constructed a strong wooden platform. These vehicles are driven miles into the forest, down tracks roughly cleared by manual brushing. The driver is accompanied by one or two assistants, who winch the logs on to the back of the lorry. Their only tools are a Trewhella (or Tire-fort) hand-operated winch and a log jack. The logs are carried to the roadside, where a full complement of between three and seven logs (depending on size) is reloaded for carrying to a sawmill. The designer would have solved this problem at several times the expense with a vehicle something like a Unimog!

Lacking heavy equipment, in the forest zones where tree fallows are usual, the small-scale farmer must depend on manual labour to clear his fields, possibly with the help of a power-saw operator to fell larger trees. The stumps and roots remain in the ground; light tractors are incapable of ploughing, ridging or harrowing it, especially if the soils are compacted and stony. I was once given a three-wheeled tractor to try out, by the manufacturer. It was provided with purpose designed implements, but after two or three weeks I returned my free gift as useless for my purposes. Usually, farmers tend to think in terms of minimum tillage, aided by a two-wheeled tractor controlled with handles, to drill seed, spray crops or tow a trailer. However, if farms are to become mixed farms, farmers must keep animals on their land, and if they keep animals they must live with them – that is they must live on their farms. Therefore the main need of the farmer will be not for a small tractor of limited usefulness, but for transport to and from his farm. Starting from this requirement, one could design a cheap vehicle, with cross-country ability but capable of speeds up to 80 kph on good roads, that could carry passengers or loads. It could have a power take-off for spraying or hedge-cutting equipment, or for driving a grinding mill or saw bench – though for the latter it might not be more elaborate than a jacked-up wheel. The vehicle would have front-wheel traction, which is cheaper than four-wheel drive, but gives better cross-country ability than rear-wheel traction, and be fitted with balloon tyres (rather like a *Beach buggy*). A flexible frame and suspension, and extra-low first gear, would make it suitable for farm work, so that it could perform many of the functions of a light tractor. The body would be integrally constructed from flat panels and tubing. The engine could be air-cooled, although this might give a danger of overheating for certain uses, and it would be of a make already familiar in the country. Hand retardable ignition would assist the engine at low speeds or when heavily loaded. The seats would be of tubular construction and canvas covered, with the rear seats folding away to make way for loads. It would be fitted with a complete or partial canvas roof. The cost might correspond to a year's earnings from a properly developed farm of 10 or 20 hectares. The resemblance to that French peasant's car, the *Citroen 2CV,* may not be coincidental!

In West Africa, and particularly in Nigeria, the trades of black smith, mechanic, welding, soldering and carpentry are widespread, ranging from highly skilled to rough and ready. The motor car has existed in Africa almost as long as in Europe. If a farmer designed some machinery, indigenous artisans could certainly put it together. The main problem is satisfactory alternative farming systems to peasant farming, for which to design machinery! Moreover machines are required that are capable of being understood and maintained by the farmer himself, rather than the sophisticated and complex machines now usual in developed countries which rely on the availability of specialised spare parts and on the help of specially trained mechanics. In Europe, the agrarian revolution was not so much a matter of new machinery – the old methods continued – but rather a change in the conception of agriculture and in farming systems. Make agriculture sufficiently profitable, and the machines may well invent themselves!

The risk of forming a 'plough pan' in the soil, is less with draught animals than with tractors; it can be reduced by using a plough that does not turn the soil, and killing the weeds with herbicide instead of turning them under; or the pan can be broken with a soil ripper, which requires a tractor to pull it through the soil. In Europe it has long been the practice to drain fields, using underground drains about 60 cm below the surface. Formerly these were ceramic pipes, or stones or brushwood tipped into a trench and covered with soil; the drains were levelled by hand. Nowadays plastic pipe is inserted directly into the ground using a powerful crawler machine. I have not heard of underground drains being used in tropical soils, nor do I know how effective they would be.

The role of smallholding

Smallholding, with farms of up to five hectares, is normally made possible only by producing high value crops such as cabbages, carrots, lettuces, melons, onions, parsley, peppers, tomatoes, etc near a town. Small farms cannot keep enough livestock to provide sufficient manure to keep the land in good heart, and need external sources of manure and fertiliser. This can be provided by town refuse and nightsoil. In West Africa, town refuse contains a high proportion of vegetable matter, especially as leaves are often used as a wrapping material. Because of his nearness to his market and lower transport costs, and minimal spoilage during transport, the smallholder can also obtain a much better margin on the sale of his produce, which compensates him for the smaller area at his disposal. He may also, for example, be able to hire out his machinery, for cultivating or cutting grass or hedges, to householders, and thus to supplement his income. Jefferies, writing in England in 1880, said, 'China, we are told, however, supports the largest population in the world in this manner. Not a particle is wasted, not a square foot of land but bears something edible. The sewage of the towns is utilised, and causes crops to spring forth; every scrap of refuse manures a garden. The Chinese have attained that ideal agriculture which puts forth the greatest amount out of it, and absolutely wastes nothing. The picture is certainly charming . . . If the agriculture of China is so perfect; if the sewage is utilised; if every man has his plot; if the population cannot possibly become too great, why on earth are the Chinese labourers so anxious to get to America or Australia, and to take the white man's wages? And is that system of agriculture so perfect? . . . Those who desire to destroy our land system should look around for a more palatable illustration than is afforded by the great Chinese problem.' He continued, 'The truth in the matter seems to be this. A labourer does very well with a garden; he can do very well, too, if he has an allotment in addition, provided it be not too far from home. Up to a quarter of an acre – in some cases half an acre – it answers, because he can cultivate it at odd times, and so receive his weekly wages (from a regular employment) without interruption. But when the plot exceeds what he can cultivate in this way – when he has to give whole weeks to it – then, of course, he forfeits every Saturday night, and soon begins to lose ground. The original garden of moderate size yielded very highly in proportion to its extent, because of the amount of labour expended on it, and because it was well manured. But three or four acres, to yield in like degree, require an amount of manure which it is quite out of a labourer's power to purchase; and he cannot keep livestock to produce it. Neither can he pay men to work for him; consequently, instead of being more highly cultivated than the large farms, such plots would not be kept so clean and free from weeds, or be so well manured and deeply ploughed as the fields of the regular agriculturist.'[7]

New farming systems

It has to be stated that there is no single right way to farm, and there is no ideal size of farm. The best way to farm is the way that best earns the farmer's livelihood and, while conserving the soil, leaves him contented. The optimum combinations of crops, and variants of

7 Jefferies, Richard (1880) *Hodge and his masters,* Chapter XXIV, *The cottage charter – Four-acre farmers*, Eyre & Spottiswood
Note: 1 hectare = 2.46 acres

livestock production, depend on local soils and climate, and on local market forces. These in turn depend on distance from the nearest town, and on access to markets, whether at home or overseas. It may also depend on whether electricity is available, on the supply of water, on what labour is obtainable locally, and on the farmer's ability to store his crops. Optimum farm size is also influenced by these factors, and will alter as economic circumstances change. As Jefferies' example shows, there are also wrong sizes of farm and wrong ways of farming. It may not be possible to increase farm size progressively from small to large, and remain economically viable across the whole spectrum. A combination of activities complicates the farmer's life, but makes him less vulnerable to the vicissitudes of a particular crop, whether from climate or market, and thereby helps average out the farmer's risk. Nor does he necessarily need practical skills other than those he already possesses; he can raise familiar crops and familiar breeds of livestock. It is clear that peasant farming no longer satisfies people.

New methods of farming can only be resolved by people trying out their ideas and modifying their techniques and the system they follow as they proceed – and then remodifying them as economic conditions change. Experimental figures are never sufficiently representative to forecast confidently how matters may work out on a practical scale; and one cannot calculate these things in theory when one is 'breaking new ground'. Of course estimates should be made first; a dead loss is likely to continue to be a dead loss in practice; but experimental results are usually too optimistic, and the amount of profit and its certainty, and the extent to which it may be improved, can only be found from practical experience.[8]

The most astonishingly eccentric systems may be highly successful. For example, an agriculturist from Sri Lanka told me that he had kept pigs, and that the effluent from the pig pens, including the manure,

8 Sumberg and Okali have recently proposed a system of small-scale sheep or goat production, combining arable cropping with growth of fodder, by alley-cropping between hedges. Two to four animals on a third of a hectare holding give an estimated financial rate of return to the farmer of 65 per cent. Unfortunately they do not appear to have included the opportunity cost of the farmer's time, and the total value of his earnings, in cash and kind, is estimated as only up to US$500; that is a third of the Government minimum wage in Nigeria. Even if farmers could each manage a hectare, it may be doubtful if their average earnings could much exceed the minimum wage.
Sumberg, J.E. and Okali, C. (1984) *Linking crop and animal production: a pilot development programme for smallholders in Southwestern Nigeria,* Journal Federal Dept, Rural Development, Vol. 1(1), 25–29

had been washed into a pond. This fertilised the water hyacinth that grew on the pond, which was fed to the pigs. There was the bonus that fish were raised in the pond, and also benefitted from the pig manure. The point is that the system is near biological balance, and that the external inputs needed can be relatively small. The same applies to the use of artificial fertilisers in mixed farming. Additions are necessary either to balance up the existing nutrient supply, or to compensate for original soil deficiencies, or to adjust soil pH. The needs are intermittent, and in their absence the system only degenerates slowly, if at all, when for example it is not possible to obtain fertiliser. Contrast this with the present situation, in which peasant farmers are encouraged to maintain soil productivity by pouring on subsidised fertilisers, and thereby to postpone the time when fallowing is needed. In the savanna zone particularly, the already low humus content of the topsoil is reduced progressively under continuous cultivation. The cohesion of the soil, already poor, is further diminished. This increases the susceptibility of the soil to damage by the erosive forces of wind and rain. Moreover, the loss of humus decreases the nutrient-holding capacity of the soil, so that fertilisers must be poured on ever more generously to maintain growth and yields. At the same time, the applications of nitrogenous fertilisers and muriate of potash (KCl) cause the soil to become increasingly acid, so that eventually it becomes impossible to grow certain crops, such as cowpea or soya. Though soil pH may be corrected by liming, lime is not readily available in many tropical countries.

Roles of husband and wife

A wife is as essential to a mixed farmer as to a peasant farmer. In Britain, a farmer's success was often attributed to his having a 'good wife' – that is, hardworking and thrifty. The varied work meant that a farmer needed help, and his children were useful too. Today in Europe, farming has become much more mechanised and specialised, and farmers' wives may lead similar lives to women in the towns. A man must live near his animals, and they must have access to his land; which means that he has to live on his farm if he is to farm efficiently.[9]

9 In England, after land consolidation and enclosure, the farmers moved out of the village to houses and farm buildings built on their farmland. However my own village enclosed in 1829, too late to benefit from the boom years of the Napoleonic wars. In fact, on their rather poor soils, it is only since Britain joined the European Common Market in 1975, that farmers became affluent enough to abandon their old buildings in the village, for modern designed farmhouses built on their own land.

Tapping a rubber tree – a job often done by women

Womenfolk need an occupation, and a way to earn money for themselves, as peasants' wives were able to do. This *modus vivendi* has to be worked out in the new system, and people can only do this for themselves, as peasant society can do. Many Africans, including farmers, are used to having an urban base, and may not live happily in 'the bush' as isolated nuclear family units – unless perhaps the price is right. It is remarkable what people will tolerate if the reward is sufficient – they may even learn to like it! Wealthier farmers require more protection than poor farmers, and law and order may need to be improved from what it has become. The farmers will want better communications, whether by road or telephone, and these should function throughout the year. However, richer farms and richer farmers will help to generate the wealth to develop rural areas. An African expects to have a working wife, and that she should contribute towards the household expenses; nor will she be happy to beg every penny from her husband. In modern industrial society, when most of the labour is increasingly being done by machines, we tend to think of equality of the sexes in employment, but in societies where manual work is more prevalent, this is a luxury, and it is more important to use the different abilities of the sexes to greater advantage – for example the greater strength of men for the heavier work, and for women the work requiring greater dexterity. There can, however, always be exceptions to prove the rule!

In plantation agriculture it is usually possible to find tasks suitable for women, or that women perform better than men do. Women can tap rubber trees, and are well suited to light factory work. They may even be happier than men doing routine or boring work, particularly if they can talk, and their minds are free to think about other things! Naturally, children are a very important part of most women's lives; this does mean that their rate of absenteeism tends to be higher than that of men, and maternity leave may also deduct from their working lives. Indeed women with small children may prefer part-time work. Nevertheless it is clearly desirable that opportunities for employment for both men and women should be similar, and that this is more likely to lead to a balanced community. The efficiency of plantation agriculture is proven for tree crops, such as oil palm and rubber, but not much effort has been made to integrate it with animal husbandry, perhaps because of the specialised training of plantation managers and agronomists. A consequence of world recession is that the world price of cash crops has slumped, especially oil palm and rubber. It is possible that the profitability of these plantations could be restored by combining them with raising cattle or sheep pastured on the cover crop beneath the trees. In any event the cover crop has to be cut back, because if it gets too luxuriant it impedes the passage through the plantations of palm bunch collectors or rubber tappers. Livestock will graze the regrowth back for nothing, and at the same time turn it into valuable meat!

New tillage methods

The writer has seen no-tillage in the International Institute of Tropical Agriculture at Ibadan, where maize has been grown for fifteen years and the grain yield has been maintained. This has been possible by using 'Paraquat' to kill weed growth, and also leaving the fallen corn stalks as a mulch on the soil surface. The new crop is drilled through the mulch. The condition and structure of the topsoil is good, stained with humus and showing intense earthworm activity. The result was impressive, but I must confess myself dubious of the universal practicability of this method in present day West Africa. Paraquat is an unpleasant chemical to use, and furthermore is imported. To what extent could a farmer feel confident of obtaining this herbicide whenever he needed it? Fertiliser additions are also necessary to maintain yields, and again they may not be available when and where they are needed. Hartmans says that no-tillage is proven superior to conventional tillage (i.e. by ploughing and harrowing), but that in both tillage systems there is a significant decline after the fourth year, to which soil compaction and soil acidification are important contributory factors.[10] Agboola comments, 'Although zero tillage [i.e. no-tillage] has been advocated . . . this cannot be practised all over the forest zone since this depends on the amount of clay, weather conditions, etc. The University of Ife has a contrasting result which indicates that zero tillage is not an ideal management practice for Ife zone. My belief is that zero tillage can be practised on sandy soils. My experience in Badeku confirms that continuous cultivation can be practised with zero tillage and minimum exposure. Most farmers in this country [Nigeria] in any case practise minimum tillage because of the type of implements they use.'[11] He refers, of course, to peasant farmers. Dr Lal informs me that mechanised drilling, spraying and harvesting result in soil compaction, and a fall in maize yields under no-till farming; but this does not happen if these operations are carried out manually. He attributes acidification of the soil to the use of fertilisers, and particularly to ammonium sulphate. He comments that, at least in the humid zone of West Africa, soils tend to be substantially sandy rather than clayey. He has observed that the use of 'Carbo-Furan' to control maize stem borer affects the soil fauna drastically, and kills the earthworms. He thinks it may be better to accept a reduction in maize yields of say five per cent, than to attempt chemical control.[12]

Live mulching is another possibility being investigated at IITA. A

10 Hartmans, E.H. (1981) *Land development and management in tropical Africa*, International Institute of Tropical Agriculture, Ibadan (p. 21)

11 Agboola, A.A. (1979) *Soils of the Nigerian forest zone and their response to cultivation,* paper in '*The Nigerian rainforest ecosystem*', edited by D.U.U. Okali, Man & Biosphere National Committee, Federal Ministry of Science and Technology, Ibadan (pp. 80–94)

12 Estimates of net income have been made for mechanised no-till maize production at the International Institute of Tropical Agriculture (IITA) at Ibadan, based on experimental findings. If land-clearing and other developmental and overhead costs are excluded, the field operations, materials and harvesting were found to total US$1045 per hectare; income, based on 5 tonnes/ha/year of grain at a market price of $400 per ton, gave a revenue of $2000 and therefore a net income of $995 per hectare. It is doubtful if consistent average yields of this level could be obtained on a commercial scale in the forest zone at present, and a half the amount could be considered satisfactory. A mechanised farming enterprise in southern Nigeria during 1984, in its third year of operation, obtained 1½ tons/ha, and with a local maize price of $600 per ton only just cleared its expenses – excluding all developmental and overhead costs. This is typical.

food crop, such as maize, is planted directly into a living cover of a low-growing plant (particularly legumes such as *Psophocarpus palustris* or *Centrosema pubescens*) with minimum soil disturbance. The live mulch system incorporates the soil conservation features of organic mulch and no-tillage, with the additional advantage of smothering weeds. The legumes contribute also to the supply of soil nitrogen. Hartmans quotes figures that suggest no diminution in yields for the main crop (maize) due to the intercropped live mulch, though intercropping usually results in at least some diminution for a particular component species. (See *Table 7*). The live mulch could be used as grazing for livestock after harvesting the main crop. There has been a long experience with green manuring using sown fallows, such as the leguminous creeper Mucuna (*Stizolobium deeringianum*),

Table 7 Effect of ground cover and fertiliser level on maize yield (tonnes/ha) (I.O. Akobunda, unpublished data)

Groundcover		**N fertiliser level (kg/ha)**			
		0	**60**	**120**	**Mean**
Conventional tillage	Early season 1979		4.37		4.37
	Late season 1979	1.48	3.40	3.11	2.54
	Early season 1980	1.24	1.49	1.51	1.41
	Late season 1980	0.67	1.69	1.31	1.22
No-tillage	Early season 1979		3.88		3.88
	Late season 1979	0.68	2.78	3.47	2.31
	Early season 1980	1.63	2.70	2.51	2.28
	Late season 1980	1.30	2.98	2.67	2.32
Live mulch (*Psophocarpus*)	Early season 1979		3.27		3.27
	Late season 1979	2.15	3.00	2.62	2.59
	Early season 1980	2.47	2.39	2.41	2.42
	Late season 1980	2.04	2.43	2.73	2.40

A late maturing maize was used for uniform first cropping, following legume cover establishment in the previous year (1978), with an early maturing maize in subsequent years.

From Hartmans, E.H. (1981) *Land development and management in Tropical Africa,* International Institute of Tropical Agriculture, Ibadan (p. 21)

which are subsequently ploughed into the ground to help maintain both soil organic matter content and soil nitrogen supply. However it has not been found as effective as natural fallows in restoring soil fertility. Moreover the luxuriant growth of the creeper can become extensively wound round the machinery when the land is ploughed. Of course an alternative is to kill the plant by spraying it with a herbicide, and leaving it as a surface mulch through which the crop is drilled.

It must be admitted that using animal manure as fertiliser is not without disadvantages. Manure can exacerbate problems with weeds, due to seeds germinating in the dung, and these may have to be ploughed under much more deeply if they are not to become a nuisance. In Europe, chemical control of weeds and the substitution of artificial fertilisers for manure have allowed people to forget how weedy fields were formerly, and what a serious problem they could become in crops. The bright colours of poppies and charlock have disappeared from English wheatfields, and the thistles out of the pasture, thanks to the intensive use of selective herbicides. In those days, weeds, pests and diseases had to be controlled by the rotation of crops. The land could be cleaned by turning it to a crop of potatoes or to temporary pasture (*ley*). After a few years, pasture would have helped restore the soil nitrogen content, and could be ploughed under, adding to the soil humus. Similarly today, the rusts on cereals or the blight of potatoes are checked by fungicide applications, whereas formerly their control depended on crop rotation, and in the case of rusts the eradication of alternate hosts. Fungicides often contain copper, and repeated applications may lead to a toxic build-up of the chemical in the soil. Could farmers in many African countries be sure that they could obtain the chemicals when they needed them? Could the countries concerned afford the foreign exchange required to purchase them, if their use became general? It is clearly better to have farming systems that are not excessively vulnerable to such considerations – and in which agricultural chemicals and artificial fertilisers are an aid to farming rather than a necessity.

Crop protection

Nematode populations build up in the soil under certain crops, and this may make it impossible to continue growing a particular crop on the same ground. Examples from the tropics are bananas and plantains, pigeon pea, peppers and tomatoes. It may be necessary to follow these by another crop, which is insensitive to nematodes, such as sorghum or groundnuts, or to leave the land fallow for a period Treatment of the soil with nematocides can be very expensive.

However some crops, such as sunflowers, are antagonistic to soil nematodes. This has also been found to be true of the invasive weed in the forest and derived savanna zones, Eupatorium (*Chromolaena odorata*), which can provide 100 per cent control of injurious soil nematodes. The weed belongs to the Compositae, as does Sunflower, and the foliage has a strong aromatic smell when the leaves are crushed. It can form thickets above head height in farm fallows, and smothers grasses including the serious rhizomatous weed, Spear grass (*Imperata cylindrica*). The sharp leaves of Spear grass can pierce yam tubers, causing them to rot in the soil. Unlike Spear grass, Eupatorium is easily uprooted, and can be burned off in the dry season, though a protective tangle of dead stalks remains. Under Eupatorium, a noticeable feature is the intense activity of earthworms. When Eupatorium foliage is cut and applied as a mulch, it disappears rapidly. Not surprisingly it is noted as a soil improver. Unfortunately the weed has no obvious use, and is unpalatable to livestock. Perhaps the solution may be to find a use for it, for example as a source of leaf protein, perhaps added to carbohydrate staple foods to improve their food value. Many useful agricultural crops appear to have begun as weeds, such as *Amaranthus* and *Celosia* in the tropics, and Mustard and Rape in temperate countries.

Control of insect pests may be possible by natural means, including sophisticated measures, such as the introduction of natural predators or parasites of the pest, in a new area, or by the release of specially bred sterile males in large numbers. However, more humble methods are also sometimes possible. For example William Page writes of the variegated grasshopper (*Zonocerus variegatus*), which can defoliate crops, 'Egg pods, which take six months to develop, are deposited in highly aggregated sites found scattered throughout cultivated areas... Egg laying sites were identified by observation and the egg pods subsequently dug up ... Counting the following season's hatching populations in the two areas showed that the potential hatches had been reduced by 83 per cent and 91 per cent respectively. Control by community self-help is suggested, where individual farmers identify and dig up laying sites on their own land, exposing the pods to desiccation on the soil surface. If this were carried out by all the farmers in an area, populations of *Zonocerus* could be reduced and crop damage minimised. Such control can be carried out at little or no cost and with very little labour.'[13] The dry-season population of

13 Page, W.W. (1978) *Destruction of eggs as an alternative to chemical control of the grasshopper pest, Zonocerus variegatus (L.) (Orthoptera: Pyrgomorphidae) in Nigeria,* Bull. Ent. Res. *68,* 575–581.

Zonocerus lays its egg pods in dense groups in the ground – and these sites may contain thousands of egg pods. The groups of insects are commonly found on shrubs and regrowth from stumps in the cultivated area, the insects laying their eggs in the shade of this vegetation.

Farmers can spot the sites by the aggregations of insects, and mark them so that they may return to dig up the egg pods. At Ibadan, egg laying occurs from mid-March to mid-April, but whereas Page recommends digging up the pods from April to October, according to Wester Modder of Ibadan University it should be done during August and September.

Crop research

The International Tropical Agricultural Research Institutes specialise in particular crops, and for this reason their investigations of farming systems tend to be restricted. If crops are to be combined together, either in space because of mixed cropping, or in time because of rotation of crops – and the system incorporates arable crops, tree crops and animal husbandry – then there must also be a flexible approach to research.

Of course, because of the constraints of staffing and space, there are limits to what a particular institute could do. The moral may be that farmers must themselves develop an experimental approach to farming, drawing on the scientific and technical advice available from agricultural research stations and university departments. This is particularly necessary for the interface, in applied research, between science and economics.

Arable crop breeding, in the tropics, often seems to be directed at producing plants for some ideal situation. For example, the strains of cassava bred for resistance to mosaic disease have proved much less resistant on the poorly fertile fields of peasant farmers. Cassava bred for high yields has been found to have inferior storing qualities and the tubers become rapidly fibrous; the custom has been for the farmer to leave the crop in the ground, if necessary for many months, and to dig it up when it is needed. Cultivars with reduced cyanide content are also more liable to rodent damage. In the case of varieties of cowpea bred to yield 60 days after sowing, the brief periods of flowering and fruiting make the plants very vulnerable to pests and diseases, so that precautionary spraying with insecticide is essential if there is to be any certainty of getting a crop; plants with more protracted growth have the opportunity to recover. The point is that perhaps crops are being bred without sufficient awareness of the conditions that they ought to be bred for – because appropriate farming systems are not yet sufficiently developed.

Up to 30 years ago, the wheats and barleys used in Europe were generally long stalked. The straw was useful, because it was used for bedding animals, and returned to the fields mixed with dung and urine as manure; it was also used as thatching for roofs and ricks, and to some extent as animal feed. However the decline in mixed farming, the general coming into use of combine harvesters, the disuse of straw for bedding cattle, the construction of government subsidised barns and the replacement of horses by motor traction in the towns caused straw to become a nuisance instead of a saleable commodity. Breeding short-strawed cereals eliminated much of the straw, with the added advantage that the crops became much less liable to **lodging** (i.e. flattening to the ground by strong winds or heavy rain). The stubble left by the combine harvester was burnt off: both to prevent a build-up of fungal diseases in the decaying straw, and to reduce the carbon addition to the soil. Recently, though, the straw is again being collected rather than burnt. This is due to new techniques of turning it into useful fodder for cattle, by treating it with ammonia or urea. Once again, in some regions, straw is becoming in short supply. Clearly the proportion of straw to grain, and their relative values and saleability, affects the profitability of the crop, and possibly whether once again mixed farming will be preferred to specialisation in either cattle or corn. Perhaps farmers in Europe will begin to want longer strawed cereals again! To breed crops for the impoverished fields of peasant farmers may seem profitless, but is it appropriate to substitute the experimental farm?

Role of livestock

Traditional livestock owners did select and cull their animals, but this was to emphasise distinctive appearance rather than to enhance their economic value and productivity. The Fulani herdsmen improved their herds in fineness of body, appearance and colour, but paid little attention to their capability for work, or yield of milk or meat. In fact these are difficult to achieve without a properly kept system of records. The introduction of improved pasture and expensive fodders means that livestock must be bred to make the best possible use of them. Animals are needed that will convert their food efficiently, mature quickly and, preferentially, develop the more desirable and expensive cuts of meat. The degree to which it is fatty or lean will depend on the food preferences of the people. In the industrial countries, individual pigs have been bred that have a conversion rate of two to one (fodder to pig live weight) though four to one is more usual, whereas for sheep it is nearer six to one and for cattle ten to one. The time taken for a beast to reach marketable size has been reduced from five or six years, before land enclosure, to about two years today. Fortunately cows bred to digest concentrates efficiently still retain their ability to do well in mixed farming systems on pasture, though of course their growth and/or milk yield is reduced.

In 1922 Lugard wrote, 'Oxen are specially valuable, since most of the tribes understand their management. They are, of course, much more useful as draught than as pack animals. The draught ox is the most valuable animal in ordinary country, the camel in loose sand, and the mule in steep country intersected by ravines. But the use of the wheel has remained unknown in Africa . . . and is only now being slowly introduced. Where the use of animals is possible, the first lesson for the African is to learn to employ the country-cart of India and China, which can be repaired in any village and requires no metalled roads, and to abandon a method of transport [head porterage] suited only to the precipitous gradients of the Himalayas. The African does not excel in the care of animals. He will kill them by overwork, and sore-back them by atrocious pack-saddles. He fails to realise . . . that an animal in hard work needs a grain ration, or that oxen, which will not graze at night, must be given adequate time to

An ox-drawn country cart

feed during the day.'[14] Here Lugard spoke of matters in which he was an expert, for he began his career as a transport officer in the Indian Army, and was selected for service in West Africa because of his skill in organising military transport under difficult conditions. Sixty years later, I have still not seen a country cart in Nigeria, though I am told they exist in Burkina Faso and have recently been introduced successfully into Bauchi State of Nigeria. Draught animals can also usefully supplement tractors, and can have the advantage of innate intelligence, for example stopping or starting at the correct moment without bidding.

Oyenuga writes, 'Domestic animals in Nigeria and West Africa have, for centuries, suffered form chronic hunger. They are poorly fed and no particular feeding regime has been evolved. The domestic sheep, goats, pigs and poultry in southern Nigeria have been living ... on waste products available from kitchens or traditional food processing industries, such as the bran from milled maize, Guinea corn or millet or the peel from cassava and yams or other root crops. Poultry are fed spasmodically with maize and Guinea corn grains ... They are left to roam about the roads or surrounding bush, picking up what they can for supplementary or main feed supplies...the animals expending, in the process of seeking feed, considerable energy which reduces from the limited food intake the amount available for productive purposes. It is not generally accepted in southern Nigeria that such animals must be regularly and properly fed to achieve normal growth, reproduction and productivity.' He continues, 'The overall result is that cattle, sheep, goats and pigs in West Africa have ... been inadequately fed. While a limited food intake may have produced hardiness in a tropical environment, it has led to the diminished growth rate, small body size and low rate of productivity ... One of the major problems facing livestock improvement in West Africa is to arrest this restricted feeding and replace it with improved and cultivated pastures during the rainy season and with supplementary feeds during the dry season.'[15]

A land-use report on the Kano area states, 'The evolution of arable farming in the intensively cultivated zone culminated in a farm [size] of about 2–5½ hectares, which is [what] an average family equipped [only] with hoes can cultivate, weed and harvest by its own [manual] labour ... But a mixed farm based on properly integrated crop and animal rotations, linked by sown pastures, which serve the double

14 Lugard, F.D. (1922) *The Dual Mandate in British Tropical Africa,* Blackwood, reissued Cass, 1965 (p. 643)
15 Oyenuga, V.A. (1967) *Agriculture in Nigeria.* FAO, Rome (p. 308)

purpose of feeding the animals and maintaining soil fertility, would require no less than 12½ hectares of arable land. Clearly, mixed farming can never be achieved on long-established farms in the close-settled zone because they are too small and cannot be amalgamated or enlarged to the required size; also because there is little prospect of diverting surplus labour from farms to industry ... For the small farmer, pasture establishment can be an expensive undertaking. Furthermore, the effort required to plant, and especially to recultivate, a temporary pasture is such that the work could not be achieved by a team of fewer than four work bulls equipped with suitable implements; this represents a considerable investment ... Integrated farming thus appears to be beyond the reach of Nigerian [peasant] farmers, at least at the present time. However it is still possible ... to introduce cattle on to small farms in the intensively cultivated areas. The animal-carrying capacity of the small farm is strictly limited by the size of the farm and by the dictates of climate. The average small farm can produce sufficient crop residues for small livestock and possibly for fattening one or two steers or alternatively for feeding a pair of working bulls through the rainy season. This means that, on a farm of less than say 8 hectares, working bulls are excluded and work will continue to be based on human labour.'

In Africa there is the need to breed into cattle the natural resistance to Trypanosomiasis (sleeping sickness) already existing in the N'dama and Muturu varieties, though tolerance to the disease is also improved by better feeding. With increased stocking rates (*stints*) of cattle, due to improved pasture and to production of fodder crops, and with increased value of the animals due to breeding, it becomes economically more worthwhile to try to control or eliminate the Tsetse fly, which spreads the disease. This may be achieved by periodically spraying with insecticide the shade trees under which flies and cattle congregate, and also by raising and releasing sterile males of the fly. Resistance to other diseases may also be fortified by breeding and by inoculation. At first, farmers may keep their own stud animals, in order to be self-reliant, but eventually they will turn to artificial insemination, and perhaps to cooperative stud farms. This can give rapid improvement in quality of livestock, and at the same time reduce the risk from sexually transmitted diseases, in particular contagious abortion.

Role of irrigation

In the arid north, the crop growing season may last only two or three months. The convention has been to extend the growing period by means of irrigation, in order both to produce more food and to provide more regular employment in agriculture. However, large irrigated schemes tend to have had a disappointing history. Despite optimistic feasibility studies, the results never seemed to justify the high capital expenditure. Moreover, the schemes may cause severe dislocations, both to human populations and to the physical environment. People may have to be moved away from areas to be flooded, and settled on new lands to be irrigated. Valuable fisheries may be destroyed and not replaced, at least for many years. Bilharzia (*Schistosomiasis*) may become a problem among communities where it was formerly absent, due to the invasion of the snail carrier of the disease into the irrigation channels. Complex bureaucracy may be required to regulate the farming and the allocation of land, and to distribute water. It may be necessary to install pumps, either for irrigation or for drainage. Tracts of land may become damaged, due to salination of the topsoil, or to poor drainage and waterlogging. Clearly the organisation needed is beyond the powers of peasant communities; their welfare and progress may be taken out of their own hands, and placed in that of an army of officials. Large schemes may have the disadvantage of inflexible arrangements for land tenure and water distribution, or for kinds of crops grown and marketing; these eventually become anachronistic because of the inertias of scale, and because they cannot respond quickly enough to changing social or economic circumstances. Of course schemes of this complexity and expense may be something that only Government can accomplish, and can sometimes be spectacularly successful. Possibly they have their place in the scheme of things, but it should be remembered that a failing project is a personal tragedy for many people, who may have set out with such high hopes. Does Government have the right to experiment with people? Nevertheless irrigation is clearly important for promoting food production in the drier parts of Africa. Perhaps smaller schemes ought to be encouraged, or schemes that can be elaborated progressively should they meet with success. Preferably they should be kept within the management scope of ordinary people, who ought to be able to decide for themselves, and Government should then help them to succeed. A community of individuals is more likely to follow a correct course of action, than a bureaucracy taking decisions on its behalf. Then, at least, they can blame themselves for their own mistakes. Larger and more sophisticated schemes may become appropriate when the people themselves have become sufficiently sophisticated to manage them.

Sylvia Sikes writes of irrigation schemes, 'Two major problems, however, emerge as one tries to grasp the real perspectives of these developments on and around Lake Chad. The first is the question as to how viable modern technical programmes really are when they are almost entirely dependent upon the complex tools and machines of modern technology, and upon operators highly trained in specific technical skills. The second is the question as to how safe these programmes are if operated in isolation from each other within a complex, and very fragile ecosystem, of which the natural regulatory mechanisms are still largely unknown. The first of these problems is sociological . . . [One may] question the wisdom of setting up large-scale schemes dependent on the successful use and maintenance of modern machinery in remote areas. Humiliating and annoying as indigenous governments may feel it to be, it has been necessary repeatedly to employ expatriate technicians and research officers, not only to install, but also to maintain and supervise such schemes. No matter how vigorously indigenous technical trainees are encouraged and provided for, they rarely manage to filter back permanently to remote rural areas to man such schemes on a sustained basis. There are other opportunities with more attractive features that do not make such heavy demands on adaptability, inventiveness, foresight, drive, individual responsibility and the willingness to make decisions personally.

'Considered in these terms, one questions for example the wisdom of setting up large-scale wheat schemes where the harvest is entirely dependent upon a sustained water supply pumped into permanent raised channels across a stretch of semi-desert. When the pumps fail one by one, the mechanics go off sick, the spare parts do not arrive, the supervisor is on leave and the senior research officer is neither willing nor able to take the pumps to pieces himself, there is a reasonable possibility that the entire harvest may fail to mature, let alone reach its market. There are no other qualified people to be found within a radius of 300 km, even if they were willing or could spare time to help out. In any case, in a land where corruption and bribery are the most dangerous diseases known, financial safeguards are so rigid that it would not be possible to mobilise funds to employ anyone at the necessary level of qualification on a temporary basis. To attempt to run such schemes at a sustained profit is thus not only unrealistic but liable to prove self-destructive: that is, unless it is an accepted fact from the start that it must be manned by a foreign technical staff.

'When we compared the wheat scheme with a market garden-cum-farm, run and developed by an enterprising Kanuri family in an inundation area of about one square kilometre near Baga, it was readily apparent that this family was making a sustained profit, with virtually no capital outlay or overheads, and no complex modern machinery or permanent structures. The ground was tilled by hand, but the farmer hoped soon to buy a yoke of working oxen and a plough. He irrigated the land in the dry weather by means of an old-fashioned *shaduf* from a shallow groundwater well, and in the inundation and rainy periods channelled the water between the ridges ... As this farmer accumulates profits he may use them to invest in more wives and concubines – that is, an increasing potential labour force. Alternatively he may invest in more modern equipment or join a cooperative society whence he can obtain the occasional services of a tractor and driver (without depending on them), supplies of fertiliser at a reduced cost, and even the loan of a pump. He may even begin to add new crops to his farm such as wheat or rice if the soil proves suitable. His economic success will result from the fact that he is independent of the use of complex technical instruments and their operators, an imposed business hierarchy, the installation of permanent physical structures, and the need to operate on a scale, and in a manner, beyond his own capacity. Above all, the enterprise is his own, and requires no training beyond what he can obtain locally in rural advancement classes.'[16] Dr Chedli Ayari, President of the Arab Bank for Economic Development in Africa (BADEA) has said, 'Irrigation is not merely an agricultural technique, it is a form of civilisation that has to develop.[17]

From the social point of view, another solution may be to provide alternative dry season employment, as a part of farming activities. In fact this is the traditional way of tackling the problem. At this time of year men go hunting and fishing, sometimes travelling long distances for these purposes; or they may work as migrant labour in the towns or on farms in parts of the country with higher rainfall, where the rains have begun already; or they may engage themselves in various handicrafts. Road building and forestry are also jobs for the dry season. Growing firewood and building poles on woodlots on peasant farms may provide both fuel and a source of income; besides irrigated market gardening, raising high value crops such as onions or tomatoes, as some have always done traditionally. Because of the smaller amount of water needed, these may be watered from wells and boreholes, but need a high input of labour. Perhaps the country should exploit rain-fed agriculture fully, before turning to elaborate and expensive irrigated schemes.

16 Sikes, Sylvia K. (1972) *Lake Chad*, Eyre Methuen (p. 234)
17 Ayari, Chedli (1985), quoted in *West Africa* No. 3526, 25 March 1985 (p. 579)

Importance of 'mixed farming'

The new commercial farming systems must be resilient and, like biological systems, 'buffered' against their environment; that is, they ought to be relatively insensitive, so that large changes in the environment produce only small changes in the system. They have to withstand whatever the climate or the economy can throw at them. Farmers will invest money in farm buildings, storage facilities, improved pasture, fencing, irrigation and drainage, tree planting, and above all in soil improvement; they will be market oriented and will use efficient methods of accounting and record keeping to assist them in management. Their farming practices will be ecologically sound and near biological balance. Contrast this with current ideas on modern agriculture in Africa – dependent on costly mechanisation of farming operations with sophisticated equipment, generous application of artificial fertilisers, and the use of expensively imported chemical sprays and dressings to keep weeds, pests and diseases at bay. No doubt, as knowledge and experience are acquired, and as systems are refined, there will be increasing specialisation and 'industrialisation' of agriculture, as there has been in Europe; but this can only occur as the infrastructure comes into being that can support this kind of agriculture. This will only happen as the country itself becomes sufficiently industrialised and developed. Until this has taken place, farmers must be as self-reliant as possible. Industrial farming is only widely practicable in societies where what I may call the 'industrial ethic' has been evolved. The sorts of social cooperation, organisation, discipline and dexterity needed are insufficiently widespread in peasant societies – which African societies substantially still are. Though peasants had virtually disappeared from England by 1900, they continued to exist in many European countries until after the Second World War, and in some places until today. They had the same ramshackle attitudes to precision engineering; in these regions the public utilities, such as electricity, telephone and water, do not have the same apparently effortless efficiency that they have in industrialised societies. I remember an elderly farmer (in England) nearly fifty years ago, boasting to my father that he would not mend a gate until it dropped off the hinges! Self-reliant farming means that farming systems must be as near biological balance as possible, so that the external inputs needed can be few and intermittent. Animals have to be incorporated in the system to make it profitable, but the farm should itself raise the fodder to feed them. At the same time there should be enough livestock to produce sufficient manure to fertilise the fields adequately.

'Mixed farming', in which the different branches of agriculture are integrated together, is a necessary phase, even though it may be an intermediate one. Unfortunately in West Africa at present, these terms are used for systems that are nothing of the sort. 'Mixed farming' is used only to indicate employment of draught animals – fodder crops are not raised, and the oxen still find forage as best they can; there is no proper integration of livestock and arable cropping. Even where peasant farmers have been encouraged to raise, by means of alley cropping, fodder crops which are cut and carried to the animals, these are still allowed to roam, and the dung and urine are not returned to the fields. 'Integrated' development applies not to the farming system, but to the provision of a full range of services to the farmer: advisory services, fertilisers, seed, crop drying and storage, marketing facilities, road communications, power, water, community welfare and so on. What we mean by *integration* is the incorporation within the farming system of arable cropping, livestock, forestry and possibly wildlife. The assembling of elaborate systems, heavily dependent on expensive inputs by Government, with all the bureaucratic effort which this entails, for the purpose of enabling illiterate or semi-literate peasant farmers to produce more food, either will fail when the support is withdrawn, or will be maintained at enormous cost out of all proportion to the achievement. It is impossible for unsophisticated rural communities to maintain such systems for themselves, however stark and simple they may sometimes seem to a European. The introduction of draught animals, and of simple but sufficiently powerful tractors, is an essential step; but it is only a first step.

It may be asked: how do nomadic herdsman, such as the Fulani, fit into these systems? The answer has always been clear – like the pygmies of the equatorial forests, their way of life is eventually doomed. Its passing may be regretted by anthropologists; but only in moments of nostalgia by their descendants. Dr Johnson wrote, somewhat fancifully, concerning the Amerindians, 'The nude Indian supplies himself with all the necessaries of life . . . as soon as his parents have nursed him up to strength, he is to provide by his own labour for his own support. His first care is to find a sharp flint among the rocks; with this he undertakes to fell the trees of the forest, he shapes his bow, heads his arrows, builds his cottage, and hollows his canoe . . . he is sheltered from the storms, he is fortified against beasts of prey, he is enabled to pursue the fish of the sea, and the deer of the mountains; and . . . he does not know, does not envy the happiness of polished nations . . . This picture of a savage life, if it shows how much individuals may perform, shows likewise how much society is to be desired: though the perseverence ... of the Indian excite our admıration, they nevertheless cannot procure him the conveniences which are enjoyed by the vagrant beggar of a civilised country: he hunts like a wild beast to satisfy his hunger; and when he lies down to rest after a successful chase, cannot pronounce himself secure against the danger of perishing in a few days; he is, perhaps, content with his condition, because he knows not that a better is attainable by man ... [but] when a tempest compels him to lie starving in his hut, he cannot justly be concluded equally happy with those whom art has exempted from the power of chance, and who may make the foregoing year provide for the following.'[18]

The nomadic peoples have, do and will fit themselves into new systems. The Fulani, when given the opportunity, have shown themselves able to take full advantage of modern education, and perform at least as well as their fellow men. For the pygmies, it is quite probably the same. Fulani have settled of their own accord in many parts of West Africa, probably when the loss of their herds through epidemic disease or drought has compelled them to turn their hands to arable farming. Settled Fulani in Gombe were among the first to use draught oxen for ploughing, because of their traditional skills with both cattle and cultivation, which they were readily able to combine together. It should not be difficult of them to develop into genuine mixed farmers, unless plentiful land and subsidised fertilisers encourage them into more extravagant methods of farming. The Fulani are fortunate in that they have a valuable asset – cows – and there is nothing to prevent them (except possibly legislation) from converting some of this asset into land. They may go through an uncomfortable transition period, and though they may lose some of their romance, they are likely to emerge from the metamorphosis better off than they were before they went into it.

I think the role of Government is not to force people into the mould of its own preconceived notions, but perhaps to observe and hasten the adaptations that people are already making in response to new circumstances, and to help them understand what is happening. Farming cannot really progress without modifications to land tenure, but these ought to be allowed to evolve naturally. By means of various fiscal measures, Government can hasten the change, for example by taxing land, and only allowing people to claim ownership of land for which they pay tax – whether the land is owned individually or

18 Johnson, Samuel (1753) Essay in *The Adventurer*, issue No. 67, Tuesday, 26 June 1753

communally. This tax should become progressively more onerous, the larger the area owned. This would encourage communities to subdivide land among their individual members, and favour the small farmer in relation to the large landowner. The payment of tax would compel people to *use* land, and discourage, but not eliminate, land speculation.

It has been shown that, in Africa, there has been almost universally a trend towards freehold land. The need is for Government to facilitate rather than obstruct this change. A legal framework can be provided, by means of which communal owners may subdivide the land amongst themselves. The law should ensure that the community carries it out with the fullest consideration of all the factors involved, that the partition is done with justice and equity, and that afterwards the agreement has the force of law. The change should be compulsory only when a sufficient proportion of members desire the change. This could be on the basis of a petition by members of the community who could reasonably show that they would expect to receive, say, two-thirds of the area to be subdivided. It is necessary, at the same time, to carry out consolidation of holdings, and planning and demarcating roads to give access to the various plots. This will be necessary, as the different freeholders will not have rights of way across each other's land after the settlement. The law should be so designed that eventually the whole nation will arrive at the same system of tenure for land ownership.

The conveyancing of land should become quick, easy and cheap. It should become the duty of Government to register changes in ownership, rather than the duty of an individual to succeed in registering a land transfer before its legality is recognised. To do otherwise is merely to create possibilities for civil servants to abuse their powers, or to enrich themselves, and to draw otherwise law abiding citizens into corruption.

Hunting and fishing

For traditional hunters and fishermen there may be only a restricted place. It is clear that the present unregulated situation, which was satisfactory when populations were small and communications were lacking, is leading to rapid destruction of the wildlife resource base. Game laws are more honoured in the breach than the observance. Individual land ownership, and the enforcement of laws of trespass, can help to preserve and regulate wildlife populations. However this cannot succeed without the approval and cooperation of local communities. The inevitable conclusion is that any land settlement must have the general consent of the community. As access to unnavigable waters can only be from the banks, it follows that, unless there is a deliberate easement, the fishing rights will belong to the adjacent land owners. Where an individual owns only one shore, then his right to fish would extend only half-way to the banks in the possession of other landowners, unless there has been some other agreement. For navigable waters, access can be allowed to the general public, but to protect the stock of fish (and other water creatures) it may be necessary to license fishing and hunting, which will be in the hands of a River Authority, in which those with traditional rights ought to be represented. There should be 'closed seasons' to allow undisturbed breeding. Permission to fish may be restricted to certain categories of people, and controlled also by adjusting the cost of a permit and the period for which it is valid.

Fish farming may only be possible in enclosed waters. It can be incorporated into farming systems: for example paddy rice and fish. Fish may also be raised in ponds supplying water power, domestic water supply or irrigation water. Several species of fish may be used to advantage: a surface feeder, such as Tilapia, a bottom feeder such as Carp, and a predator such as Snakehead or Nile Perch to reduce the numbers of small fish. Some of the fish species farmed may be herbivores.

Fish farming may also be integrated with poultry farming or pig farming, the poultry or pig manure being used to fertilise the pond. Excessive growth of waterweed, which might otherwise choke the pond, can be used as stock feed and as a green manure to put on the land. Fish offal can be dried and ground, to be used in feedstuffs for livestock.

With individual proprietorship of the land, hunting can only be permitted to those who own the land, and to those to whom they give permission to hunt on their land. For others, the only access for hunting will be to publicly owned or vacant lands. In the case of Forest Reserves in anglophone Africa, the right to hunt was usually given only to those who possessed the right at the time of constituting the reserve. It is therefore presumed that they may continue to exercise the right, and it may be desirable that they should assist in the protection and regulation of wildlife. Thought may also have to be given as to whether the communities with hunting rights may alienate them, by selling them or offering them for rent, or, in the event of their abandoning them, to whom they should pass. In the event of any general land settlement, consideration ought to be given to these matters, and to conditions which may be imposed at the time of such settlement.

Vacant lands and forestry

Vacant lands could be defined as unused lands for which the community, or other individuals, do not wish to pay tax. The landowning community may either sell them, or subdivide them among their members who may retain them, sell them or rent them. From the time of coming into possession of the land, the new owners must, of course, pay land tax. Where the tax is not paid, the Government may distrain on the land to recover the unpaid tax, returning any balance to the land owner. Of course, the land owner may sell part of his land, or all of it, or raise a mortgage, in order to pay the tax. However it is not intended that land tax should be particularly onerous, at least initially. It could be treated as an alternative to income tax for those farming land. The tax, paid over a period of time, can be regarded as a proof of land ownership. Vacant lands may be expropriated by Government, subject to a public notice, and a right of appeal. Provided that vacant lands are not under notice of expropriation, they are available for incorporation into the lands of the community to whom they are regarded as traditionally belonging, subject to commencement of liability to pay tax. However all lands claimed as individually owned ought to be subject to tax.

Forest reserves should be exempt from any tax on land, but should remain vested in the communal land owners. This should however be subject to the reserve being under a management plan, approved by the Head of the National Forest Service. The national Government ought also to consider paying an annual bounty to the communal land owners, in proportion to the area of the reserve, also on condition that it is under an approved plan, correctly implemented and monitored by the Forestry Service. This might serve as an inducement to proper mangement, and also encourage the reservation of vacant lands, on which the community is reluctant to pay tax, but which, at the same time, it does not wish to expose to the risk of expropriation without compensation. It is essential that the priority of the forest reserves is to grow wood, and that arable crop production within forest reserves is secondary to tree establishment. Other tree crops, such as cocoa, coffee, rubber or oil palm, should not be planted within forest reserves. Because they are permanent crops, the writer's experience is that this inevitably results in alienation of the land from forestry use. In reserving vacant lands three categories may be recognised.

1. Production reserves, which form the permanent forest estate, to supply timber and wood products to the nation in perpetuity
2. Protection reserves, where it is proposed to maintain the natural vegetation cover on watersheds, mountainous regions and unstable

terrain, or areas of special biological interest

3 National lands, scheduled for development other than forestry, but meanwhile placed in the charge of the Forestry Department to ensure controlled exploitation

Legislation may be needed to enable vacant lands to be treated in this way, and to define the degree to which existing forest laws should apply to the different categories. The dedication of a permanent forest estate is essential, both to provide the long term security needed for investment on a large scale in forest industry, in order to prevent the fragmentation of the estate into small uneconomic pieces, and also to allow investment in forest regeneration by giving the assurance that the new crop will have the opportunity to reach maturity. Nevertheless, there is a tendency throughout the tropical world, to spend money on doubtful methods of regeneration, when the greatest priority is the surveillance and protection of the forest. The urgent needs are as follows:

- to maintain clear demarcation of the boundaries of the forest estate on the ground;
- to strengthen the protection and patrolling staff;
- to improve Forestry Department discipline and efficiency;
- to reforest, with timber crop plantations, areas under threat from encroachment;
- to prepare and implement forest management plans to integrate the different objectives for a reserve, as regards the Nation, the State or Province, and the local communities, in order to ensure sustained or improving yields;
- to conduct public education, both nationally and locally, to explain the importance of forest reserves;
- to carry out sociological research to determine the wants and aspirations of rural communities, and how they may be satisfied.

In Nigeria, the existing arrangement for the administration and management of forest reserves is highly unsatisfactory. At present these seem, in effect, to have been expropriated from the communal land owners by State governments, and are managed neither in the national interest nor that of the local communities. Because of a worsening shortage of wood supply, it is becoming apparent that a considerable proportion of the natural vegetation within forest reserves must be reforested with more productive, and more valuable, intensive timber crop plantations. The original forest has to be exploited as thoroughly as possible, and the remainder clear felled and burnt, with a view to establishing a new crop. However it is also possible to underbrush natural forest, and poison the overwood (which is most effective with sodium arsenite – formerly used on a considerable scale in Nigeria for forest improvement), and then to underplant with fast growing species such as *Cedrela odorata, Gmelina arborea, Terminalia ivorensis* or *T. superba,* and possibly with *Triplochiton scleroxylon.* These species grow well on leached soils under higher rainfall. Teak (*Tectona grandis*) grows best on drier base-rich soils, planted into arable cultivation (*taungya*). On deep sandy soils, mixtures of Mahoganies may be planted with a 'nurse crop' of Opepe (*Nauclea diderrichii*), the latter being exploited for transmission poles. In savanna areas, and also on highly leached soils in the forest zone, it is possible to plant tropical pines, such as *Pinus caribaea,* and, in the drier savannas, *Pinus oocarpa.* The Eucalypts, such as *Eucalyptus camaldulensis, E. robusta, E. tereticornis, E. torelliana* and *E. citriodora* are also suitable for savanna afforestation. The leaves of the last named yield oil of citronella. *Eucalyptus deglupta, E. torelliana* and *E. urophylla* can be planted in forest areas.

While establishing tree crops, it is possible to raise arable crops or pasture, utilising the residual fertility from destroying the natural vegetation. Government Forest Departments have carried out successfully the experimental and pilot stages of afforestation, but are proving incapable of the large-scale planting now needed. My opinion is that, for this to be accomplished, it will be necessary to turn to commercial management companies. If this is begun by Government within forest reserves, then the expertise will also become available for planting on private land (as is at present done in Britain).

In fact, most of West Africa is not particularly well endowed with wood. High forest covered a relatively small proportion of the total area. Natural forest has evolved only to exist, and not to produce timber sufficient to support numerous human populations at a high standard of living. Moreover, natural forest is a complex mixture of different sizes of trees belonging to more than a hundred species within a particular locality. The volume of wood ranges between about 200 m^3 and 500 m^3 per hectare, of which perhaps 100 m^3 is utilisable for timber, but less than 50 m^3 is actually used. The savanna woodlands are much more extensive, but have evolved rather to resist the rigours of climate and the ravages of fire, than to produce high growth rates or good quality timber. Even for firewood, the natural regrowth cannot keep pace with present demands, especially those of urban societies. Consequently West Africa is in increasing danger of exhausting its wood supplies. The forest reserves in Nigeria are inadequate to furnish even the existing needs of the nation. But communities have become very reluctant to entrust lands to Government for forestry development, even where sufficient vacant lands are available for this purpose. The inevitable conclusion is that timber crops must also be planted on lands outside forest reserves, and this means planting by private individuals and organisations. In the pursuit of self-dependence, farmers must plant their own trees, both as hedges and as tree lots. They should be encouraged by whatever fiscal measures are practicable, to make plantations to produce timber, poles and fuelwood for their own needs and for those of urban communities, in both the high forest and the savanna zones. However, perennial crops cannot be a secure investment, particularly in the savanna zones, unless land tenure problems are resolved.

Land inheritance and farm size

Inheritance of freehold land should evolve progressively to a common system throughout the country. The method of inheritance can be a potent factor preventing the accumulation of too much land in too few hands. Inheritance can be based on the traditional African system. That is, in a polygamous household, an equal division for each wife, then divided equally among the children of that wife. A wife only inherits land from her husband in trust for her children. A childless wife also receives her share, but only in trust, and on her death it will pass in due proportions to the heirs of her deceased husband. An acknowledged child of a man by a woman other than a wife (i.e. an illegitimate child) receives a share equal to the average portion received by the other children. As is already the case in various parts of Africa, including Nigeria, sons and daughters ought to inherit equally, and this should be made a general rule. When, during his lifetime, a man indicates a preference before disinterested witnesses that his property be divided equally among all his children, regardless of the identity of their particular mothers, this should be done. A woman's property should only be inherited by her own children, and shared equally between them regardless of their paternity. Legally adopted children inherit only from their adoptive parents and not from their natural parents; they ought to inherit equally with any natural children of their adoptive parents. Where they are adopted by only one parent, they would receive an equal share with the other children of that parent, as for an illegitimate child. The real estate, that is the land, should be divided into lots of proportionate value among the heirs, unless the estate is below a certain area (say ten hectares) when it may be bequeathed to one child who must compensate the other children (selling land to do so, if need be). Above this area it must be divided, unless some children (having attained their majority) voluntarily accept compensation in lieu of their share. Of course, where a wife survives, during her lifetime the land may be retained intact, and be managed by her – the income being hers – but the land should be subdivided on her death among the children; or alternatively during her lifetime if she may so elect.

In the case of a childless wife, her personal estate reverts to her own family. For a polygamous household, trustees should be appointed to distribute the land, and these may be appointed by the bequeather during his lifetime; or otherwise by the courts. Moveable property, on the other hand, should be inherited by the children in the same proportions, though individual items may be bequeathed according to the wishes of the deceased. Of course a person may disinherit an unfilial child, but this could be disputed in the courts, and either confirmed or rejected, or the child allowed a reduced inheritance, depending on the actual merits of the case. An individual, being of sound mind, ought to be able to make gifts of personal property during his own lifetime, to other individuals. However, this would be subject to a capital transfer tax due to Government. On death, capital transfer tax could apply to any estate above a certain area of land (say 500 hectares) or above a certain value. These, or similar measures, can influence the size of farms.

The role of government

The fiscal measures of Government can also have a profound effect on farm size, and on the efficiency and profitableness of farming, and on progress in the development of improved farming methods. It has already been explained how the imposition of a graduated land tax may stimulate the productive use of land, discourage the accumulation of large estates, help establish individual ownership of land, and counteract speculation in land. It may also help to promote owner cultivation, which has time and time again, in country after country, demonstrated itself to be the most efficient method of farming. If the prices of farm produce become too low, the capital needed will not be invested in farming. Development of farming will be impeded, or entirely discouraged; at worst people will abandon farming as a livelihood – or they may only produce enough food to supply their own minimal requirements. Due to the speculative nature of commercial farming in West Africa, because sufficiently profitable and secure farming systems scarcely exist, anyone investing in farming will want the prospect of a much higher return on capital invested than might otherwise be the case. This means protection, and perhaps financial assistance, to farming. African Governments have tended to favour the consumers rather than the producers, and sometimes only to represent themselves! This has contributed substantially to the present parlous situation in agriculture. A policy of cheap food has also encouraged the drift to the towns. Price restrictions have kept the earnings of farmers far too low, and government Marketing Boards (or parastatal equivalents) have compounded the difficulties by their inefficient or incompetent operations. Instead of being paid in cash, farmers may have received chits that they have had difficulty redeeming; their produce may have rotted at the roadside due to difficulties in moving it, or rotted in storage due to inadequate storage conditions. The situation has been made worse by official rates of exchange which may be several, or even many, times greater than the unofficial market value of the currency, and indeed than its true value.

A consequence is that the farmer can scarcely buy anything worthwhile with his money, so that he has little incentive to produce. Meanwhile the growth in food imports results in insufficient foreign exchange to buy 'luxury' goods – which eventually include such humdrum commodities as tinned milk, kerosene, matches, soap, tins of sardines etc – not to mention television sets, transistor radios, record players or iron beds. Motor transport becomes difficult because tyres, spare parts, engine oil and even petrol become scarce, or even unavailable. Finally the whole trading system begins to break down. Then the country is indeed on the slippery slope to disaster. Corrupt and selfish politicians or civil servants may hasten the march to destruction. Import licences become virtually a licence to print money, as the goods are bought at the official rate of exchange and sold at the unofficial. Of course such favours require reciprocal favours, and this fosters bribery and corruption amongst Government officials. Worse still, emerging industries in the country are ruined by the unfair competition, and the essential growth of industry is sapped at its base. Productive activity in industry, and the employment which it generates, are essential corollaries to agricultural development. Meanwhile there are vast armies of civil servants to be fed, whose salaries have swallowed up so much of the Government budget that too little remains actually to enable them to do any work! The efficiency of Government consequently deteriorates still further. The paint begins to peel from the formerly splendid edifices of Government buildings, often erected with funds milked from agriculture.

Eventually 'Humpty Dumpty falls off the wall', but unfortunately, all the king's horses and all the king's men cannot put Humpty together again. Foreign investors stay away, because the prospects of ever getting their money back are too bleak. Indeed the decayed infrastructure of power, water supply, roads and transport may make it almost impossible to run any viable business at all! The country becomes dependent for mere survival on grudging handouts of foreign aid. A few countries have almost reached this terminal condition. Colonisation originally enabled a foreign power, and its nationals, to invest with confidence in regions where business prospects might otherwise have seemed uninviting. The imperial power established and maintained the infrastructure and conditions that were necessary for successful commerce and development. But who would recolonise these countries today? Indeed is there anyone that one would like to be colonised by? Today, populations are ten times what they were at the time of the 'Scramble for Africa' by the Imperial powers, and the problems are correspondingly more intractable.

In Western Europe, customs barriers against imports, and taxation relief, subsidies and grants have had a remarkable effect on stimulating agricultural production. In England, agricultural production has more than tripled in fifty years. Now the pendulum may have swung too far in the other direction. The consequences are gross overproduction of certain commodities, such as dairy products, cereals, beef etc, together with a wasteful use of resources. Livestock are fed expensive concentrates to increase their growth and production, while their dung and urine are washed away into sewers, streams or rivers to pollute the water supplies. Meanwhile cheap fertilisers are poured on the land to grow cereals, and are leached into the river drainage systems causing eutrophication, killing fish and adversely affecting potability. Agricultural chemicals, such as fungicides, insecticides and weedkillers, are sprayed and resprayed, until the toxic residues are poisoning the rivers and accumulating in the soil, in the crops, in the livestock, in the drainage water, and eventually in human beings. Marginal lands, which might be better afforested or left as rough grazing, are brought under the plough. Because of the thin topsoil and relative fragility, they become liable to serious erosion or structural deterioration. Of course, this extravagance has to be paid for, and the expense is met by industry. A consequence is that Western Europe has become less competitive in world markets.

High interest rates result from this profligacy, in order to retain capital in the country; this discourages investment, and industry is unable to borrow the money to modernise, to expand, or to risk debt. Marginal businesses are forced to liquidate, or to go bankrupt, perhaps bringing other companies down with them. During the resulting recession, at a time of reduced overseas earnings, Third World countries suffer mounting debt on borrowings made during the previous more buoyant period, further exacerbated by the rising interest rates. Perhaps this is the price being paid to keep alive peasant survivors and redundant industries in some European countries? There should be moderation in all things.

There are three vital functions of Government: to maintain a sound currency, to keep law and order and to uphold national defence. Perhaps a fourth ought to be added: national communications; it is possible for individuals to provide more or less everything else for themselves, but as governments have concerned themselves with furnishing more and more services to the community – education at all levels, health, social welfare, public libraries, inspectorates of this that and the other – bureaucratic indigestion has set in. The efficiency of the four essential services has declined along with everything else, until the consequences have become painfully obvious.

In the former colonial territories, though not in the metropolitan territories, professional officers ran the professional departments with a high degree of delegation of powers to field officers. Administrative and clerical personnel did not account for more than ten per cent of the staff of professional departments. In Nigeria, in 1959, the total number of senior civil service employees (that is equivalent to the commissioned and warrant officer ranks of the army), of the Federal and Regional Governments combined, did not exceed 5000, to govern a country (then) with a population of rather more than 50 million. At least half the officers were of Nigerian nationality at that time. Today the number of people in executive and professional grades (grade level 08 and above) of the civil services of the Federal and State Governments must be more than 100 000, to administer a population of nearly 100 million.

The adoption of the bureaucratic systems of the former metropolitan countries – dominated by administrative and clerical personnel, which may number more than half even of professional departments – has reduced the effectiveness of Government almost to vanishing point.

Mellanby writes of the University of Ibadan in 1958, 'In Britain between six and twelve per cent of the income of universities is spent on administration, and the newer and smaller the institution the higher is this proportion. We were only spending some eight per cent on this head . . . I was often criticised because at Ibadan our administration was understaffed (on looking back I do not believe this was true, for I believe with Professor Northcote Parkinson that administrative staff make work for each other, and when an administrator complains of overwork the number of his subordinates should be reduced not increased)...'[19] I doubt if British universities are now as efficient in administration as they were then. It is possible that in Nigerian universities today, administration and clerical services account for nearly half the numbers employed. It could almost be said that the sheer size and inertia of the civil service has become an obstacle to economic progress. The cost has outrun the ability of agriculture and industry to support it. In developed countries, the civil services maintain an illusion of efficiency; but the reality may be very different. As the civil service grows, and as the administrative personnel grow within it, a powerful vested interest is created that can be very difficult to resist. The people can lose control of their own government, which begins to exist only to look after itself, and to suck the wealth out of the productive part of the economy. Perhaps the growing prodigality of governments throughout the world has precipitated a 'crisis of government' which is not without some bearing on the world's present economic difficulties.

In Nigeria, massive oil revenues have inevitably given Government an important role and considerable influence in national development by the way that it disburses its funds. In most anglophone African countries, the inhabitants will accept US dollars, sterling, or francs in exchange for their own currency, at a fraction of the official rate of exchange. In the cities, mains electricity may be off more often than it is on, water flows only intermittently from the taps, telephones are more ornamental than useful, rubbish begins to block the streets, and the tarmacadam is corroded by potholes. Essential commodities are bought only with difficulty or at considerable expense. Theft and armed robbery have become almost unbridled. Government departments are expanded in the vain hope that the newcomers may do the work that the present incumbents are neglecting to do; new government organisations are created in the expectation that they may achieve what the established departments are failing to accomplish; merely to add more 'white elephants' to the existing herd of white elephants. The developed countries are suffering the same malaise, but not quite to the same desperate degree. It is people who solve problems, not governments. And it is people who will solve the problems of agriculture.

The problem of traditional farming is that its flexibility is too limited; but there is no virtue in replacing it with systems that will themselves prove restrictive, and unable to evolve. For example, farm

19 Mellanby, Kenneth (1958) *The birth of Nigeria's university,* Ibadan University Press (p. 263)

size must be able to adjust to the optimum for different cropping systems, and to changes in market forces. Moreover, as national standards of living rise, the earnings of farmers must rise too. This means progressively larger farms worked by fewer hands, together with increased profitability per hectare, which can only be achieved by greater mechanisation of farming. Mechanical aids must be devised that suit West African conditions, and avoid compacting the subsoil or causing loss of the topsoil. Certainly the biological constraints are harder to master than those which confronted the farmer in Europe; but knowledge and technical capability are already sufficient. Unlike the Europeans, who emigrated to America and Australia, Africans cannot escape the problem, or the blunderings of their own societies, by fleeing to the wide open spaces of less inhabited continents. It is in the ghettos of the industrial countries that the impoverished refugees from the underdeveloped nations will try to seek their salvation. Given the industriousness and enterprise of many African peoples, and given a modicum of competent government, the problems can be resolved. Even in Europe the metamorphosis has been relatively recent, brought about by population growth and industrialisation. England was the first country to leap the hurdle, and then only in the first half of the last century. Different localities in England spread the change over more than a hundred years. Just as they crossed the threshold in ragged order, so it is for the rest of mankind.

If political doctrine is preferred to the lessons from two centuries of human experience, if timidity or apathy inhibits decisive action, if selfish and corrupt government goes uncurbed, if people are not allowed and encouraged to solve their own problems, then in Africa we may see poverty-stricken populations dependent on a declining resource to feed themselves. Then, surely, the race will be to the swift, and the battle to the strong.

7 The implications for education

> *'There is no doubt that conditions would be better if growth and development [of education] could go together. One reason why this has not happened is that we have put the cart of educational development before the horse. Instead of starting with "What type of education is best for Nigeria", we should start in fact with the question: "What type of nation should Nigeria be?".'*
>
> *Pius Obanya*

'The mental disease of the present generation, is impatience of study, contempt of the great masters of ancient wisdom, and a disposition to rely wholly upon unassisted genius and natural sagacity. The wits of these happy days have discovered a way to fame, which the dull caution of our laborious ancestors durst never attempt; they cut the knots of sophistry which it was formerly the business of years to untie, solve difficulties by sudden irradiations of intelligence, and comprehend long processes of argument by immediate intuition . . . It is however certain that no estimate is more in danger of erroneous calculations than those by which a man computes the force of his own genius.' Samuel Johnson continued, 'The man whose genius qualifies him for great undertakings, must at least be content to learn from books the present state of human knowledge; that he may not ascribe to himself the invention of arts generally known; weary his attention with experiments of which the event has been long registered; and waste, in attempts which have already succeeded or miscarried, that time which might have been spent with usefulness and honour upon new undertakings.' But he then wrote, 'No man ever yet became great by imitation. Whatever hopes for the veneration of mankind must have invention in the design or the execution; either the effect itself must be new, or the means by which it is produced. Either truths hitherto unknown must be discovered, or those which are already known enforced by stronger evidence, facilitated by clearer method, or elucidated by brighter illustrations . . . Fame cannot spread wide or endure long that is not rooted in nature, and manured by art.[1]

1 Johnson, Samuel (1751) Essay from *The Rambler* No. 154 of 7 September, 1751

Lugard commented, 'Each generation is advancing intellectually beyond its predecessor, so that the younger men view with increasing impatience the habits, traditions and ideas of their elders.' He continued, 'The influence of education for good depends on the qualities of character it is able to evoke.' Also, 'The object which education in Africa should have in view must be to fit the ordinary individual to fill a useful part in his environment, with happiness to himself, and to ensure that the exceptional individual shall use his abilities for the advancement of the community and not to its detriment.'[2]

Virtually every country in Africa now has at least one university. The first university in tropical Africa was Fourah Bay College (of Durham University) in Sierra Leone, founded in 1876. The University College of the Gold Coast (Ghana), and the University College of Ibadan in Nigeria, were both founded in 1948, under the aegis of the University of London, and awarded University of London degrees. In 1965 Ibadan became a university in its own right, awarding its own degrees. There are now 23 universities in Nigeria, of which ten have faculties of agriculture, and five have faculties of veterinary medicine. Until recently, the only forestry department was at Ibadan, but at least three other Nigerian universities have begun forestry courses, and eleven have been proposed. The Ibadan Department has split into a Department of Forest Resources Management, and a Department of Wildlife and Fisheries.

Because agricultural production in Nigeria has remained mostly in the hands of peasant farmers, the main avenues for employment of university graduates in agriculture have been in government extension services to farmers, in research institutes and in teaching agriculture, rather than in farming itself. Thus the direct impact of higher education, especially on arable crop production, has been very limited.

In forestry, graduates are mainly employed in government forestry services; in the past they were mainly concerned with revenue collection from forest exploitation, protection and management of forest reserves, and research activities, rather than in large-scale reforestation. However, unlike most agriculturists, forestry graduates have had the executive experience of running a national forest estate. The effectiveness of the veterinary services has been somewhat intermediate in that, although the ownership of livestock has remained largely in traditional hands, they have helped boost cattle

2 Lugard, F.D. (1922) *The Dual Mandate in British Tropical Africa*, Blackwood, reissued Cass, 1965 (p. 643)

populations by the control of epidemic diseases. The value of an individual cow has made this worthwhile. Though most livestock are still maintained in traditional ways, poultry are an exception, and perhaps have been the first commercial success story in agriculture, due mainly to the participation of 'business' men and women. If Africa is to continue to feed, fuel and house its increasing populations, then the future must lie in development of agri-business rather than, as in the past, in forest conservation and in farming extension to peasants.

Existing university courses are not producing graduates with training particularly suited to the task, and there have been difficulties in finding suitable indigenous personnel for agricultural projects, and perhaps also of carrying them out adequately. A report comments that the content of agricultural courses is not sufficiently related to subsequent employment needs. Syllabuses have ill-defined objectives, and course content is often too theoretical and lacks required competences. There is insufficient effort to relate training to West African conditions and to the wide ecological range in the country. The rapid expansion in student numbers (tripled in a decade in Nigeria) has hampered the ability of agriculture faculties to maintain the quality of teaching. Equipment deficiencies are serious and widespread – almost half the major items are not functioning – but as numbers of post-graduate students increase, the need for more complex equipment will also grow.

Poulsen writes, 'From a situation where agronomists, livestock experts and foresters often look after their own narrowly defined land-use interest, competing more often than collaborating, it is time that we started to evolve toward a new situation characterised by informed collaboration leading to a high degree of integration of the basic land uses.'[3] He is, I suppose, thinking of integration at the Government level, and particularly in land-use planning. My own thesis is that it should be integrated, so far as is possible, within the same individual – the farmer. This is a route to robust, remunerative farming. However, at present, forestry, veterinary, wildlife and range management are generally separated from agriculture, which includes livestock and orchard crops, sometimes in separate ministries. The educational aspects are also separated off under the Ministry of Education. Similar schisms are found within universities, with the different subject areas either in separate departments or even in separate faculties.

3 Poulsen, Gunnar (1978) *Man and tree in tropical Africa,* IDRC, Canada

In Europe, the increased specialisation and industrialisation of farm production has emphasised the divisions in agriculture. In the tropics, plantation agriculture, of oil palm and rubber, has tended to be industrially based and also to be specialised. Faults of former colonial administrations, and of the European education system, were that they encouraged rivalry rather than partnership between the different branches of agriculture. However, if it is accepted that, in Africa, the growing of staple foodstuffs and of timber crops can only become sufficiently profitable when they are combined, either with each other and/or with animal husbandry, and that farming must become more integrated, then teaching of agriculture must also become more integrated.

Rural education

Progress in farming depends on the production of educated farmers and an educated rural population. Success in agricultural cooperation, which can do much to promote profitability and stability in agricultural markets, also depends on education, and on inculcating in the population a strong sense of social responsibility. This can be both done, and undone, at school. For two decades, West Africans have seen the apparent success of individuals who have been noteworthy for their selfishness, dishonesty, lack of patriotism and corrupt dealing. This has brought these men great wealth and their countries to the edge of ruin. Adedeji writes, 'Nigeria has to face most seriously and urgently the dangers which corruption and corrupt practices pose. Whereas a public official may stand condemned nationally and even be convicted in a court of law for corrupt practices, he may still be regarded as a hero in his primordial society which sees nothing wrong in his robbing the "alien" civic public. How to bring about a consensus of values is consequently one of the greatest challenges of our time for without it we cannot hope to eradicate corruption.' Nigeria is not alone amongst developing nations in suffering this problem.[4]

Plato, quoting Socrates, wrote in about 390BC concerning primary education, 'And the beginning, as you know, is always the most important part, especially in dealing with anything young and tender. That is the time when the character is being moulded and easily takes

4 Adedeji, Adebayo (1985) *Nigeria - Lessons of experience, West Africa* No 3560, 18 November 1985 (pp. 2417-2419)

any impress one may wish to stamp on it . . . Then shall we simply allow our children to listen to any stories that anyone happens to make up, and so receive into their minds ideas often the very opposite of those we shall think they ought to have when they are grown up?'[5] In 1922 Lugard, advocating a code of honour, wrote that we should, 'place before children in an attractive way, by illustration and anecdotal biography, the social and other incentives to gentlemanly conduct, and the success which rewards honesty, self control and industry. Truthfulness, courage, fair-play, respect for authority, good manners etc are treated from the secular point of view of their value in social intercourse, and their results in obtaining success in life'.[6] The principles of education, which he urged, were formerly inculcated in the government colleges (for secondary education) in Nigeria. Judging from personal acquaintances, these colleges were formerly remarkably successful in instilling a strong sense of social responsibility in their alumni. In England, since the introduction of the comprehensive school system after the Second World War, schools have less and less regarded themselves as responsible for the moral instruction and discipline of their pupils. Quite certainly the result has been a moral deterioration of the nation – by no means disastrous, but evident nevertheless.[7] This may be a modern reaction against the Victorian belief in 'original sin', and their obsession with rehabilitating the degraded inhabitants of urban slums, and creating a ruling class to administer a far-flung empire – it definitely succeeded in generating, in the general population, a strong sense of duty to the community. This is expressed by the sentiments of those, from all classes of society, who went to the slaughter of the First World War. Perhaps two World Wars have helped reduce enthusiasm for simple patriotism. However, the revival of Africa may depend on this inspiration.

Obanya, writing of Nigeria, says, 'Schools have increased in numbers but is is doubtful if they are preparing their products for life. There is no doubt that conditions would be better if growth and development could go together. One reason why this has not happened is that we have put the cart of educational development before the horse. Instead of starting with "What type of education is best for Nigeria", we should start in fact with the question: "What type of nation should Nigeria be?". Our answers to such a question

5 Cornford, F.M. (1941) Translation of *The Republic of Plato,* Oxford University Press (p. 366)

6 Lugard, F.D. (1922) *op. cit.*

7 During January 1986, a BBC programme stated that in Great Britain a third of young males had received a criminal conviction.

should involve the clear definition of a Nigerian approach to development, a national cultural policy, technological/scientific policy, and a rethinking of an educational policy which will have very close links with national goals and aspirations.'[8]

Large-scale mechanised State farms have failed, both commercially and technically, and the onus of feeding the nation must therefore continue to rest on 'private enterprise' farming. Throughout the world, the experience is that owner-cultivation is the most efficient and productive way of farming, and this must be the goal. Peasant farming is, of course, private enterprise farming; but it cannot be made sufficiently remunerative to satisfy a more educated population, nor justify the expense of their education and training. A source of discontent at government trade schools was formerly that those qualifying were unable to obtain earnings commensurate with their skills, however hard they worked – unless they were employed by Government. Lugard commented, 'Though little material reduction in the cost of education can be expected from school fees, it is, I think, important that they should be imposed, however nominal in amount. The African is not singular in regarding as of little value what costs him nothing.'[9]

Because of the need to develop new farming systems, farmers will need to be innovative. They will require a thorough scientific understanding of their environment, to be trained as soil managers, instead of seeing themselves as soil exploiters, to understand simple economics and financial control, and to be able to do bookkeeping. Their training must embrace arable cropping, animal husbandry, forestry and possibly wildlife management, whatever is appropriate to their locality. Many farmers, though, will have neither the opportunity nor the ability to go to agricultural college or university.

Dr Arthur Porter, former vice-chancellor of the University of Sierra Leone, has said, 'One of the major constraints to the development of agriculture, and the linkage of the modern sector of the economy with the rural, has been the low level of basic education – particularly in the rural areas. This made it difficult for farmers to take full advantage of the investments being made in the sector through the provision of credit, the supply of seeds and the facilities for marketing. It is clear that . . . improvement of basic education in the rural areas, to ensure a strong foundation for development, must be a priority.' He said also,

8 Obanya, P. (1983) *Growth and development in Nigerian education since Independence*, paper at the National Conference on Nigeria since Independence, Ahmadu Bello University, Zaria, March 1983
9 Lugard, F.D. (1922) *op. cit.*

‘There appears to be little or nothing in the system which prepares its recipients for self-employment or to occupy risk-taking, decision-making positions in the economy. Instead, rigid adherence to the passing of examinations as the criterion of success implies conformity to work habits rather than scope for individual initiative. More particularly, the present educational system, to the extent that it reaches the rural areas, is not related to the circumstances of rural life patterns.’ He explained, ‘This misfit, or lack of sufficient relevance between what is provided and what is needed, is true not only at the primary level but throughout the system. It certainly is true for technical education. In many of our countries the courses provided are those intended to turn out Western-style artisans, able to fit into a skilled work-force as urban wage-earners, even though an insufficient number of such jobs is available; while what is needed is a steady supply of self-employed artisans with some business knowledge, able to seek out and develop for themselves business opportunities related in the first instance to agriculture.’[10]

Primary education

Primary education normally lasts six years, up to the age of eleven or twelve. Because of the expense, at least for several decades, primary education will be the entire education that the great majority of the population will receive; but progress in both agriculture and industry requires an educated population – as does progress in democracy! Primary education is therefore preparing people for life, as well as some pupils for entry to higher education. It must give a thorough grounding in the ‘three Rs’: **R**eading, w**R**iting and a**R**ithmetic, so that at a pinch people can continue to educate themselves afterwards.

Primary education necessarily begins in the local vernacular language; but a start should be made in whatever international language is appropriate in the territory concerned – whether English, French, Portuguese or Spanish. This opens up a considerable range of literature, both technical and general, which is otherwise denied. Africa will need farmers who are literate and numerate, as well as practical, and who will continue to broaden and extend their knowledge by reading.

It is important to imbue an innovative approach. This may be achieved most simply and cheaply by teaching *applied biology* as a

10 Porter, Arthur (1985) in a paper delivered at the conference on *Education priorities in Sub-Saharan Africa,* Windsor, England, quoted in *West Africa* No 3549 of 2 September 1985

practical experimental subject based on the school garden. It may only be feasible in rural areas, where sufficient land is available, though pot experiments may be possible for town schools, as they require less space. This is intended to give the children both the concepts of 'chance', and of improvement by experiment – because in a traditional African community, nothing is regarded as happening by chance! The pupils can measure natural objects, including their own heights, to obtain an idea of 'normal distribution' and of the difference between two 'populations' (for example boys and girls); though it is realised that algebra and statistical analysis are beyond children at this stage. The importance of repetitions, or 'replications' can be explained, demonstrated by simple fertiliser and manuring trials. The children can observe the growth and life histories of crop plants, livestock, insect pests, fungal parasites, and of course of human beings. Elementary genetics can be explained and demonstrated. The course can be linked to other courses: language to write essays and reports; mathematics to calculate the results; economic geography and so forth. Such courses would need to be illustrated by well prepared textbooks, because at first the teachers may not be much more skilful than their pupils. The teachers can motivate the children to review an area of study, to pose the questions, and then to seek the answers. Imaginative free work on the part of classes should be encouraged, for this is the basis of innovation.

Cooperative studies could be included with the teaching of history and geography, explaining its origins, its development, and the social responsibilities and duties necessary for success. The children should always be shown the example of previous generations, and the satisfactions and rewards of true merit. The rural areas and rural towns are starved of reading matter. Perhaps subsidised literature could be provided in the vernacular, with cartoon stories for the semi-literate, designed in simple language to interest and inform the reader. Television can be an excellent popular educator, producing great improvement in children's ability to understand English (or whatever is the metropolitan language), as they watch programmes in English, where otherwise they rarely hear English at home.

Secondary and post-primary education

At secondary school, it is essential to give future agricultural students a thorough grounding in science subjects. This is particularly necessary if the sixth-form and A-levels, where the student spends two years in a class with about twenty others, is replaced by one year of 'preliminary science' at university in a class of about 200. (At Ibadan

University the preliminary class for chemistry numbers nearly 1000.) Chemistry, biology and ecology should be emphasised. Aspects of agriculture could be integrated into an 'applied ecology' course, with a high experimental content. Again this may be based on the school garden, and use readily obtainable materials. This may help solve the problems schools have to obtain or afford the experimental equipment and materials needed for the more academic subjects. Elementary courses in conservation, including forest conservation, wildlife conservation, and soil conservation could be included. Prospective agricultural students also need to be taught geography because of its links with geology, land-use and economics. Francophone countries seem to fare better than anglophone in giving a good foundation in the metropolitan language and in mathematics.

Pupils normally leave school after either five or six years of secondary education, at about 16 or 17 years of age, after the 'School Certificate' or 'O-level' examinations, or after seven years, at about 18 or 19 years, after the 'Higher School Certificate' or 'A-level' examinations. The *Alternative Biology* (Ordinary level) syllabus of the West African Examinations Council in 1984, in many respects fills the role of an 'applied ecology' course as the writer conceives it. (See summary of the course, *Table 8*.) This particular course syllabus has not become popular, though it has been available for about a decade. It is demanding on teachers, and may be very different from what they were used to when they themselves were at school. Perhaps the component subjects ought not to be taught in the order given in the syllabus. In any event the visits and practical work must be related to the seasons of the year, and to the farming cycle. It is intended that farm plots should be cultivated and that experiments should be carried out, for example in crop rotations, use of manures and fertilisers, and so on. Peasant farming is barely mentioned, or its merits or demerits explained; nor are the advantages and disadvantages of mechanical cultivation indicated for discussion. Those who drafted the course may have assumed that the general use of fertilisers and mechanical cultivation are the obvious cures for Africa's agricultural ills. In earlier chapters it is explained that the matter is not as easy as this. If the course is not successful, perhaps it should be asked whether the fault is with the course or with the teachers – and whether and how this may be rectified. Such courses, and those in agriculture, assume that land is available for farming at the school. This may or may not be true, and there could be difficulty teaching the course at town schools; in Nigeria a substantial proportion of the population now lives in cities. Students proceed from this course into the normal Biology (Advanced level) course.

Table 8 Alternative Biology (Ordinary Level)

A summary of the 1984 syllabuses for the School Certificate and General Certificate of the West African Examinations Council. This is divided into six main topics.

1 Agriculture

A Farming

(i) types and methods of farming; cultivation; irrigation

(ii) maintenance of soil fertility

B Livestock

— establishment, feeding and care of animals e.g. fowl, pigs and cows.

C Fisheries and fish ponds

— establishment and maintenance

D Economic importance of agriculture

(i) Food crops and cash crops

— their production, uses and economic returns

(ii) Diseases of plants and animals

— of cocoa, rubber, palm, groundnuts, tobacco, corn and cotton

— trypanosomiasis in cattle; coccidiosis in poultry and guinea pigs

(iii) Food problems

— storage, pests, transportation, population control

2 Interactions in nature

A Ecological methods, instruments and tools

— simple measurements; collection of specimens; collection and interpretation of data

B Basic ecological concepts

— population, community, ecosystem, habitat, dominance

C Succession in a habitat

— study all the year round of aquatic *or* terrestrial *or* arboreal

D Energy relationships of living and non-living things

— energy flow; food webs; nitrogen, carbon and water cycles

E Population attributes

— density, fluctuations, territories, growth etc

F Interrelationships among living things

(i) commensalism, cooperation and mutualism

(ii) competition, predation and parasitism

G Applied ecology

— implications for agriculture, conservation, management of land and wildlife.

3 Man and microbes

A Microbes on living organisms

(i) useful microbes

(ii) harmful microbes

B Sanitation

(i) organisations *(ii)* social responsibilities

(iii) water purification *(iv)* refuse disposal

(v) sewage disposal *(vi)* pollution

4 The cell

A The cell as a unit of life, freeliving unit, part of living whole

B Structure and function of the cell

(i) cell membrance *(ii)* cytoplasm *(iii)* nucleus

C Processes in living systems

(i) diffusion *(ii)* osmosis *(iii)* plasmolysis

D Cellular metabolism

(i) anabolism *(ii)* catabolism

5 Structure and physiology of organisms

A Transport systems (circulatory systems etc)

B Control of cell environment – roles of kidney, liver, skin, lungs

C Coordination of the cell environment

(i) irritability – stimuli and response

(ii) chemical coordinations – hormones and auxins

(iii) skeleton in relation to movement

6 Sexual reproduction, variation and adaptation

A Sexual reproduction – in plants and animals

B Optimum conditions for seed germination, and development in animals

C Organisms in populations – variations and normal distribution

D Adaptation and genetic selection

E Control of characters

(i) family characters and elementary principles of heredity

(ii) selective breeding

I have visited a progressive grammar school in the west of Nigeria, and was profoundly impressed. Under the guidance of their headmaster the students helped build their own classrooms, operated a school printing press, constructed a dam and thereby an irrigation/swimming pool where they learned to swim, planted orchards, ploughed and cultivated food crops, kept cows and poultry, and ran and maintained the school tractor. The headmaster contributed considerably from his own pocket towards these expenses. The practical instruction, outdoor exercise, character formation, discipline, self-reliance and amusement given to the boys was considerable. Though there can be no objection to teaching agriculture as a school subject, to those who wish or can be induced to learn it, the cost and effort involved tends to be high, and perhaps out of proportion to that needed for other subjects. Modern agriculture is a highly capital-intensive business, and schools are not usually the sorts of institutions that can run it profitably.

It must be recognised that few secondary school leavers with agricultural qualifications will want to work in peasant scale farming. They will hope to continue to a Government Technical School, or to study agriculture at university, and later will either enter the Civil Service, or abandon agriculture altogether. The considerable capital required by modern agriculture, simply could not be entrusted to a young man straight from school or university. Indeed, at present, it would be difficult for him to get the relevant experience after leaving college. Even in Europe, for a young man to enter farming, he must either have a rich father, or one who owns plenty of land, or preferably both! Many students in later life may wish to take up gardening, but it is a healthy pastime rather than an economic activity, though nevertheless it can help men to feed their families and to enjoy themselves at the same time.

I confess to being sceptical of the merit of introducing wholesale teaching of agriculture at the primary and secondary school levels as a routine subject in school curricula. It is apparent that it would boost the employment opportunities for agricultural graduates to teach the subject, but would it do much to enhance national agricultural production? It might be a wasteful use of resources and a diffusion of effort. How can one teach economic farming, that one does not know how to do, to children who do not want to learn it? Could schools afford to buy the farming equipment, or succeed in maintaining it, or make sufficient use of it to justify the expense? Livestock cannot go into suspended animation during school vacations.

Agriculture may be taught in special District Agricultural Colleges at post-primary level. Here students learn the nature of, cultivation and management of the soil, the control of pests and diseases attacking farm crops, tree planting, grafting and nursery work, and the use of machines and machine maintenance. They could receive additional grounding in basic sciences, English and mathematics, and be taught simple economics and account keeping. The students would be drawn from the district in which the college is situated, so that the instruction is relevant to the environmental, sociological and economic needs of the particular area. It must be emphasised, though, that college leavers will be very dissatisfied with a training that does not afterwards recompense them for the time and expense of their education. Initially such a programme ought only to be on a pilot scale, to see how it turns out. The production of large numbers of college leavers ill-fitted to the present economic situation may be undesirable, and cause serious social problems.

The 6:3:3:4 system

In Nigeria, the 6:3:3:4 educational system is being introduced; that is six years primary school, three years junior and three years senior secondary school, then four years at university – though enthusiasm for the change varies between different States. The hope is to promote a system more 'relevant' to Nigeria's needs, leading into technical education as well as into the more academic paths that dominated the former system. The details do not seem to have been entirely worked out, but the intention appears to be to make secondary and further education available to a large proportion of children. During the 1970s, Nigeria had the feeling of being endowed with unlimited wealth, and could contemplate a 'democratisation' of education, with secondary and higher education available to all who wished to benefit from them. This may have led to favouring higher education at the expense of primary education. Yet primary education is generally not of the same quality that it was twenty-five years ago, when a 'Standard 6' school leaver might be more literate and numerate than many an 'O-level' candidate today. The cost is, perhaps, beyond any but the rich technically advanced countries and, in effect, attempts to construct an imposing edifice on weak foundations. Moreover, such systems may be undesirable, by giving people expectations that cannot be realised – if anything adding to the country's problems by creating a dissatisfied and alienated section of society who have little to contribute to economic or social development. An incidental consequence of the new system is that, compared with the former 'A-level' system, many young people will go a year earlier to university, where they are less supervised than at school; this is a source of anxiety, at least to the parents of daughters.

New systems of education can easily create social problems worse than those they were intended to cure. In England, during the 1950s, the comprehensive school was introduced to replace the secondary modern school and the grammar school. This was despite the fact that the grammar school was proving highly successful at producing the technocratic elite whom Britain so badly needed, and was open to all children from whatever social background. The change was intended to solve the problems of catering for both the non-academic and the academic child at the secondary level, consequent on the raising of the school leaving age to 15 years, which had occurred in 1947. The secondary modern school was regarded as stigmatising the less intelligent child, and putting the late developer at a disadvantage. Middle class parents, who could not afford private education, were not content for their children to receive an apparently inferior education at a secondary modern school. The desire was to replace a meritocratic system with one that was more egalitarian, and less obviously discriminated between children. The comprehensive school lumped all children together, whatever their innate intelligence or background. These schools were large, sometimes catering for 3000 pupils. It was anticipated that this would make possible a more efficient use of teachers and teaching facilities, and offer to each child a greater range of teaching options than could be catered for at smaller schools. It also, incidentally, made possible more elaborate hierarchies of teachers' responsibilities, increasing numbers of administrative posts and thereby enhancing the pay prospects of senior teachers.

After 1964, the Government attempted, with substantial success, to enforce comprehensivisation of remaining grammar schools, in order to prevent many of the more able children being siphoned out of the comprehensive system. Moreover, the grammar schools had represented particularly the middle classes, who took an intelligent interest in their children's school, and in the academic progress of their children, which had an important influence in motivating their school work. In many comprehensive schools, though, this influence became swamped. Another consequence, because of the large schools, was a deterioration in the personal relationship between teacher and child. It may be that these have led, ultimately, to football hooliganism and drug abuse by young people; and as undisciplined children are more likely to become undisciplined adults, eventually to the overflowing jails! However an important contributing factor to the loss of school discipline was probably the raising of the school leaving age itself, eventually to 16 years in 1972. Older children are more difficult to discipline than younger children, and increasing the compulsory school age resulted in frustration for the non-academic child. The lack of discipline in the senior classes spread back into the younger age grades.

Attempts have been made to reduce the size of comprehensive schools, by splitting them into junior and senior schools, or by separating off potential university entrants in 'sixth form colleges' for the last two years of secondary education, to prepare for the 'A-level' examinations and university entrance. It has been alleged that this has exchanged a system of not knowing the children because there were too many of them, for another where one does not know them because they are not at the school long enough! The generally unsatisfactory performance of many comprehensive schools has, if anything, strengthened the position of the 'public' (i.e. private!) schools, by removing their only serious competitor – the grammar school. If egalitarianism grinds everyone down to the lowest common denominator, then it may spell mediocrity and in the end a mediocre nation. Egalitarianism is a luxury.

People from developing countries tend to look at advanced countries, which they wish to emulate, and to think that their education systems have always been the way they are now. As stressed elsewhere in this book, their present systems are the result of development and not the cause of it. They are often quite inappropriate in developing countries, and not necessarily particularly well suited in the advanced country itself. Changes in the education system may have all sorts of ramifying social effects, further modified by the ways children and parents respond to them, that can be difficult to perceive when they are first introduced. It is best to modify systems in ways that permit a retreat to first base, in case the change should prove disadvantageous; and if possible to dip in a toe before taking the plunge. These matters ought to be considered when redesigning educational systems.

Technical education

Most African governments have long experience of providing technical instruction in agriculture, forestry and veterinary studies at an intermediate level (i.e. post-secondary 'O-level'). These schools are often administered within the Ministry of Agriculture, rather than the Ministry of Education. The courses are usually for two years, and originally students received either bursaries or salaries from Government for the duration of their studies. These schools gave specialist technical training for students who eventually joined the Government agriculture, forestry or veterinary services as technicians. The courses were later recognised for entry to university, giving exemption from the initial year of preliminary science; as did also the successful completion of sixth-form 'A-level' studies at secondary school. These technical schools usually have competitive entrance examinations, and attract able students, formerly because they virtually guaranteed employment by Government on successful completion of the courses and also because they became a route to university for students from poorer homes, who could not afford the expense of sixth-form studies at school or a preliminary year at university. Sometimes students were even able to contribute to their home budget from their bursaries. In Nigeria today, the number of self-financed students at these schools is increasing, and they already make up more than two-thirds of the total intake at the Forestry School at Ibadan; the situation is similar at the other technical schools. For them, there is no bespoke employment at the end of the course, and they must compete for Government employment afterwards. In addition, former students who have spent several years working in the field since qualifying, are taken back for a further two year course, during which they are funded by their employers.

The technical schools prepare for the following examinations, or their equivalents.

- An initial two year course leading to the Ordinary National Diploma (OND). This is essentially a practical course, for example giving training in surveying, methods of cultivation, use of fertilisers, husbandry of livestock and poultry, diseases of crops and livestock and their treatment and control, methods of farming extension, use and maintenance of machines, etc. The students also receive a grounding in science subjects, and in English and mathematics. They will perform and report on class projects. They may be better taught, have better supported teaching, and there are often fewer in a class than for university students. The general experience has been that when these students go on to university they perform in the top half of their classes, and make above average employees afterwards.
- A subsequent two year course, after qualifying in the initial course followed by several years working experience. This leads to the Higher National Diploma (HND), which is recognised as equivalent to a university pass degree. The students receive additional practical training and experience, and will receive instruction also in man management, project accounting and financial control, and the more academic aspects of their specialisations which would form a part of university courses, such as farming systems and farm management, range management, economics etc. Each student also carries out a special study (i.e. a personal investigation) usually of a practical nature, and prepares a report on the results.

There are still problems in accepting students from this level on to university postgraduate courses. In part this is because of university regulations and perhaps an over-zealous protection of 'university standards'. There may also be some jealousy between the university and the upstart polytechnic systems (see below). It is recommended that these students should be accepted at university for a year's 'Diploma' course, aimed to fill in deficiencies in their former training *vis-à-vis* university courses; those that perform sufficiently well in this course should be allowed a further year's M.Sc. studies, and if found to be merited, to proceed to a Ph.D. degree. It is best if the technical schools continue to be run by the Ministry of Agriculture (or whatever is the appropriate ministry) rather than by the Ministry of Education, or by the Ministry of Science and Technology if this includes the Government research establishments. The reason is that otherwise they become too 'boffin-oriented' instead of being practically oriented, and begin to worship excessively at the academic altar. They were originally intended to produce staff trained in specific skills required by Government technical departments. When control is lost by their own department, this purpose tends to be lost sight of, despite carefully composed Boards of Control. The graduates become less practical and give less satisfaction in their working duties – or at least that is the danger. Though successful candidates from the technical schools were originally all earmarked for Government jobs, some pass into employment by universities, and by commercial organisations. This is a good thing, and should not be discouraged. The students should not be bonded by Government, unless the drain on personnel becomes more than Government can afford. One presumes that a loss of up to fifty per cent of the students following graduation could be tolerated, and ought to be very beneficial for national development.

In Nigeria, polytechnics have been established with the hope of supplying technically trained personnel for commerce and industry. A 'foundation course' may be provided, giving more intensive teaching to get students through 'A-levels'. There is a tendency for polytechnics to become 'sub-universities' (a trend seen also in Britain). Every polytechnic instructor appears to regard himself as a university lecturer *manqué*. It is doubtful whether the polytechnics are as effective as the government technical schools in their particular specialities. This is partly because they cater for those who do not qualify for university entrance, partly because of larger classes, and partly because they do not have the supporting staff and facilities of a government department. At polytechnics, students qualify with OND and HND qualifications, and perhaps eventually with B.Sc. or B.Eng. and M.Sc. or M.Eng. degrees. A wider range of instruction is offered than at government technical schools. It is apparent that their spheres overlap, and that at some stage the two channels will have to be harmonised and merged. However a consequence of this will be to transfer all technical training to the Ministry of Education. This has been commented on already. Perhaps government technical schools should continue to function as they are, so long as they can cope with the entire national demand for technicians of a particular category, for both government and non-government needs.

It is important that polytechnics and technical schools should recruit a good proportion of their instructors from industry and the ministries, with substantial practical experience, and maintain strong links with these other institutions. They should not, as universities tend to do, indulge in the incestuous habit of keeping on their own students to join the teaching staff. When these institutions have transferred to the Ministry of Education, the tendency is to insist on academic excellence for appointment and promotion of staff. In consequence teaching becomes less and less related to practical situations and needs. A report commented that training objectives are often unclear, and that insufficient importance is given to motivating the students with correct attitudes; course content is dictated by traditionally designed syllabuses, in which the relationship between theoretical and practical aspects is insufficiently emphasised. The report urged a national system of accreditation, for syllabuses and examination standards, a determination of priorities, a reduction in the diversity between institutions, and a more equitable distribution of resources among them.

Government established 'cooperative colleges' serve mainly to train students to work as cooperative inspectors in Government service. These operate in the same way as the government technical schools, though they fall within the Ministry of Cooperatives. Only a small proportion of students pass out to work in the cooperative movement itself. It is apparent that if cooperative colleges are to be run for the benefit of the cooperative movement, they should be funded and administered by the cooperative movement itself, though perhaps with the assistance of Government grants, and with Government representatives present on the Boards of Control. Such colleges could accept post-primary, as well as post-secondary students, the respective courses being geared to their educational levels. Students could be on bursaries awarded by individual cooperative societies, or from a joint fund, or subsidised by Government, or any combination of these. The colleges would also give additional training to officers of societies, for example in financial control, accounting, business administration, personnel management, public relations and so on. They may also be able to organise instruction in machine use and maintenance for the more important industries in which cooperation is engaged. There may be a need for the colleges to run courses to publicise the 'cooperative ethic' to different interest groups, such as schools, universities, civil servants, politicians, journalists and other interested parties.

It is important, if the cooperative movement is ever to stand on its own feet, that it should have a position separate from that of Government, and this should include cooperative education as well. However, the ability to set up independent colleges is contingent on the financial strength of the cooperative movement. Government grants-in-aid are not objectionable, but if the Government contribution is excessive, then inevitably he who pays the piper calls the tune. It is not desirable that the cooperative colleges should become either academic institutions or the agents of Government, but their courses should be relevant to practical cooperation, and to what the societies need and want.

University education

The reasons for the relative lack of direct impact of university education in agriculture on progress in farming have been referred to in the introduction to this chapter, and in the first chapter of the book. The point has been made that, if agriculture is to become more integrated, so must the teaching of agriculture, and that students in the different fields of agriculture must be brought up to regard each other as collaborators rather than as rivals, as in the past. It is neither possible, nor necessary, within the scope of this book, to indicate in detail the syllabus of a university course. Nor indeed can this be done for polytechnics, nor for technical schools, nor for secondary nor primary schools. This task is better served by a series of committees than by an individual author, but it is possible to indicate a line of thought along which university courses could be organised – to try to achieve the purposes that have been outlined. Variations may exist between universities, depending on local circumstances, on the qualifications and experience of the staff, as well as on national needs. Because universities have internal examinations, though standards may be monitored by a system of external examiners from other universities, it is possible for each university to have its own particular theme. This specialisation depends on the climate, geology and land-use characteristics of the country, or region, in which the university is situated; and may also be influenced by the particular interests and qualifications of the staff that it appoints. I will take as an example a system intended for an agriculture faculty at a university in a high rainfall part of the forest zone near the coast.

This zone is most suited to growing perennial crops, such as bananas, coconut, oil palm, raphia palm, rubber, timber and pulpwood crops; cocoa may also be grown in one locality on volcanic soils, though the region generally is not suitable for cocoa. The needs of tree crops have aspects in common, and would benefit greatly from the pooling of expertise – though the usual practice has been to teach forestry crops separately from agricultural tree crops. In this particular region there are extensive vacant lands available for plantation agriculture. Dumont writes, 'Modern policy towards forest land is rapidly becoming intensive and productivist, instead of limiting itself to the sterile framework of protection and conservation ... Like cultivated plants, the planted forests of tomorrow will benefit from the development of good species, fertilisation, irrigation and plant health measures. The twenty-first century will thus see forestry schools closing down, because forest cultivation will be taught along with orchard cultivation, in agricultural institutes. I hope that Africa does not copy our current arbitrary division of disciplines, which are already out-dated without our realising it.'[11]

The zone is not suited to pasturing cattle, as the soil tends to leach and deteriorate under grass cover. However livestock could graze on the leguminous ground cover normally sown under palm or rubber plantations. It is impossible to maintain the fertility of these soils under arable cultivation for more than about two years, though the period may be protracted by using fertilisers and green manures. However arable crops, such as coco-yam (taro), egusi melon, maize, cassava, pineapple or yams can be raised during the period of establishing the perennial crop. The added cultivation also benefits the growth of the trees. Alley cropping may be possible, with zero or minimum tillage, growing maize or cassava between the hedgerows, which are cut to provide foliage to stall feed goats or other livestock. Tree foliage is usually more nutritious than grass in the humid tropics. It might also be possible to keep dairy cows in screened cattle sheds, with periodic spraying against insect-borne diseases. Poultry and pigs may also be kept, perhaps in conjunction with fishponds. Market gardening would be possible near the town, using urban refuse and artificial fertilisers as manure. There is a brewery in the city, and the brewer's grains and other brewery waste could be used to stall feed

11 Dumont, René (1966) *False start in Africa,* Andre Deutsch

dairy cattle, Freshwater, lagoon and marine fishing is also possible within the region. Wildlife cannot be the tourist attraction that it may be in the savanna game reserves, but there could be opportunities for birdwatching along the coastline and around the lagoons. Considerable areas are still forested, and there are also mangrove swamps and freshwater forest swamps. There is therefore a role for forest conservation.

It was proposed that the theme for agricultural specialisation at this university should be *tree crops*. Specialisation towards particular tree crops would only be required in the final year of the course. Nevertheless graduates would have a more open field for employment, whatever their particular specialisation, and particularly for plantation crops in both agriculture and forestry. The Agriculture Faculty would be split into seven sub-departments:

- agricultural economics;
- animal science;
- arable crops;
- horticulture;
- rural sociology and extension;
- soil science;
- tree crops and forestry.

I am not in favour of including comprehensive courses in 'Wood Science and Technology' in the programme of what is essentially a biological department of agriculture. This ought to be taught in the Engineering School of the University, along with Agricultural Engineering. The reasons are that expensive mechanical equipment and machinery are needed if these courses are to be taught adequately. It can be difficult for forestry graduates to find jobs in the wood manufacturing industries, but by including other engineering subjects their prospects for employment would be greatly improved. Nevertheless it is not possible to work entirely in water-tight compartments, even for faculties; and whether or not it is envisaged to include wood technology amongst engineering subjects, it will be essential to teach conversion and end-uses of wood to future forestry and tree crop plantation managers. They are likely to become involved in forest clearing, and perhaps in production and conversion of estate timber for fencing, buildings etc as well as for sale. Moreover they will require training in surveying, road-making, bridging, elementary building construction, machine maintenance, irrigation and drainage.

The scope for practising 'Wildlife and Range Management' in the region of the university is limited or non-existent. This would not merit any considerable input of funds, as it might in a region of extensive savanna forest reserves. There is little pressing need for public investment in this domain, and little possibility of recouping the expenditure by culling wildlife populations, or through the tourist trade. Nevertheless wildlife conservation should be included in land-use courses. In these circumstances, it may be sensible to include range management in Animal Sciences, together with animal husbandry, fisheries management, and animal biochemistry, physiology and nutrition, where it may more properly belong. Fisheries are so important in this region that, perhaps, they ought to be split off as a separate faculty, though some instruction in fish farming will be necessary for agriculturists. All these subjects should have strong linkages with Zoology in the Science Faculty.

Assuming students enter the university with only an O-level or 'Brevet' qualification, for a five year course, in the first year it is essential that they should receive a good grounding in general science subjects, English and mathematics. For reasons of economy, they may share courses with students entering the Biological Sciences Faculty, though if classes become too large the two streams could be separated. In designing courses, the intention is that a student may delay choosing his particular specialisation as long as possible. This gives greater flexibility to his choice of career; and such a system is more responsive to rapidly changing regional and national needs. Students often arrive at university deficient in English and mathematics, and need remedial teaching in these subjects, if they are to perform well in subsequent science courses.

Unfortunately, over-large university classes often make it difficult for students to receive the personal attention that many of them require. At the end of the first year, the weaker students may be weeded out. It is a waste of resources, as well as damaging to an individual's self confidence, to allow him or her to continue with a study programme that can lead ultimately only to failure.

There are several subjects that are needed by most students across the spectrum of agriculture. These include meteorology and hydrology, soil science, crop botany, conservation, statistical methods and experimentation, data processing and computer use. Therefore these would be taught as a *common trunk* for all agriculture and forestry students during the second and third years. This consists of *introductory courses* for each of the seven main subject areas (c.f. departments), and also a series of *basic courses* (see *Table 9*). All these courses would be obligatory. Clearly, for some of the basic courses, teaching staff would have to be borrowed from other faculties – though course content should be slanted to agriculturists' needs. Accounting includes financial and cost accounting; crop botany includes plant taxonomy, anatomy and physiology. I do not favour

Table 9 Courses of the common trunk in second and third years

Introductory	**Basic**
Arable cultivation Animal husbandry Horticulture Tree crops and Forestry Agricultural economics Rural sociology and extension Soil science	Accounting Data processing/computer use Statistical methods/experimentation
	Meteorology and hydrology Wood conversion and utilisation Elementary agricultural engineering and ergonomics
	Crop botany Ecology plant/animal Conservation soil/forest/wildlife Seed biology Crop protection and pathology Biochemistry and nutrition Genetics

stipulating courses in which a pass is compulsory, before a student is allowed to proceed to the next year of studies. This leads to endless pleading for exceptions. Agriculture is a sufficiently roomy discipline to accommodate a wide range of talents. Failure in a particular course only influences the special subjects for which a student is subsequently accepted; or in the event of a particular specialty having reached its quota, whether he is accepted at all. Course marks are the final arbiter between students in the competition for places, and this saves argument. Again, weak students can be weeded out at the end of the third year, based on total marks.

In the fourth and fifth years students enter their area of specialisation. This would belong to a specific department. The choice depends both on the student's preference, and on the selection made by each department. Year four concentrates on the common aspects of the specialisation areas in that Department, and also on practical field experience. For example, in 'Tree crops and Forestry', there could be:

- plantation management and management planning;
- economics of tree crops;
- tree physiology;
- tree protection and pathology;

- plant propagation and breeding;
- cultural methods and control of fertility;
- erosion control;
- multiple land use;
- statistics and biometry.

All subjects would be required. The students would also have to prepare an actual management plan for a plantation, based on a study of a particular estate on the ground, and the collection of the necessary statistical information.

The fifth year would entirely comprise elective subjects. In addition, the students would have to write a 'special project', based on personal observations or survey, or experimental work. The courses offered would all be specialised, and from them each student would make two choices (though perhaps limited by compatibility of timetables):

- banana crops;
- orchard crops, including coffee and rubber;
- palm crops;
- forest inventory and mensuration;
- forest law and administration;
- forest or plantation management;
- nursery organisation and production.

A summary of the organisation of the courses over the whole five years is given in *Table 10*.

Some of the traditional subjects, such as 'Silviculture' could be broken into their components, and taught as part of 'Plant ecology', 'Plant physiology', 'Forest conservation' and 'Forest management'. Agronomy might receive similar treatment. The different courses can be coordinated together by means of 'Statistics and Experimentation' – a central course, in which the actual experiments carried out as a part of the student's other courses and his special study can be used to provide material for statistical analysis by the student.

He should also receive practice in both sampling surveys and experimental design and analysis. My experience is that it is better to teach the practical use of statistics, before explaining the theoretical basis. Throughout the developing world, and in some industrial countries too, statistics is taught as an academic subject largely divorced from the real and practical. It is necessary that students actually do experiments and analyse them, instead of only having them described or written on the blackboard. This suggested reversal of the usual order of presenting subjects can also be applied in other fields. For example 'Animal husbandry' may be taught before 'Animal sciences'. However there could clearly be difficulties in

Table 10 Scheme for course organisation

<table>
<tr><th>Year</th><th colspan="8">Topics</th></tr>
<tr><td>1</td><td colspan="8">General science subjects (+ English and mathematics)</td></tr>
<tr><td>2</td><td>Science (3 subjects)</td><td colspan="7">General agricultural subjects</td></tr>
<tr><td>3</td><td>Science (2 subjects)</td><td colspan="7">General agricultural subjects</td></tr>
<tr><td rowspan="2">4</td><td rowspan="2">Science (1 subject)</td><td colspan="7">Special agricultural subjects</td></tr>
<tr><td>Agric. econo.</td><td>Animal science</td><td>Arable crops</td><td>Horti-culture</td><td>Rural social and ext.</td><td>Soil science</td><td>Tree crops and for.</td></tr>
<tr><td>5</td><td>Science (M.Sc.)</td><td colspan="7">Elective topics within specialisation</td></tr>
<tr><td colspan="9">Note: This assumes that undergraduate courses in Science subjects are for four years.</td></tr>
</table>

teaching 'Plant breeding' before 'Genetics', and I am reluctant to press these opinions too far outside my own experience.

Programming and computing, besides being of considerable practical use, are a valuable exercise in logical thought. Micro-computers and programmable calculators are becoming extremely cheap, and also extremely powerful. In Africa they will become the stuff of everyday life, as they are becoming in Europe. A protest by the uninformed, that they will put clerks out of work, is a misconception of their role. Computers are as likely to generate employment as to eliminate it. They have strengths and weaknesses that the human mind has not got; they make possible much that was formerly virtually impossible, or too expensive in time and money to be undertaken. Experiments can be performed and analysed that would have been impracticable previously; commercial efficiency is greatly enhanced by the degree of control and speed of monitoring industrial or financial processes. There is no resisting this advancement, if developing countries are to catch up with the developed nations. One agrees that it can be taken too far; in many developing countries it does not do to rely on excessively sophisticated methods or equipment that collapse as soon as the expatriates who have established them go away. Computers, though, are becoming increasingly robust, and a start has to be made and the relevant teaching must be given. Developing countries have the advantage that they can short-circuit some of the stages in technological progress, and cna even leap-frog developed nations in the process.

The syllabus described should be seen as indicating the sort of way in which curricula could be redesigned in agricultural faculties. The theme, and emphasis on various specialities, would vary between different universities. This would depend on local circumstances, both ecological and economic, and must also be influenced by the particular interests, and strengths and weaknesses, of the teaching staff at each institution. For example, a university in the Sudan zone would not be interested in tree crops, but in cattle, arable crops and the production of firewood. Irrigation might be important. Universities in cities located in the less populated savannas, might be more interested in cattle ranching, wildlife and conservation. In some areas peasant farmers wax strong, and are still a considerable component of society, and here farming extension would be an important consideration. There can be no single pattern; the diversity will add to the variety of agriculture graduates, and to their joint capacity to meet whatever are, and will be, the needs of agriculture across the country.

Problems of higher education in agriculture

A recent report, commenting on the teaching of agriculture in Nigerian universities, writes that farming extension is nowhere a separate department, except at Ibadan, but usually only forms a small part of a department of agricultural economics. Agricultural engineering appeared as a separate department only at Zaria (Ahmadu Bello), and as 'Agricultural Technology' at Maiduguri. Irrigation, despite its great importance in the north of Nigeria, was considered mainly as an engineering problem. Where irrigation was taught in depth, the facilities were deficient. The recent realisation of the need for a farming systems approach, was not adequately reflected in the curricula. Universities tended to fall into one of two groups:

- well established institutions which had greatly expanded enrolments with very little or no expansion of their resource base;
- new institutions with poor physical resources due to inadequate

financing when they were established.

Most agriculture faculties possessed their own farm, workshop and other facilities, but the quality varied. Some farms were only two or three hectares, whereas others extended over several hundred hectares. Most workshops were inadequately equipped and poorly maintained. Crop storage and processing facilities were often inadequate, and livestock housing variable in quality. Libraries were generally poorly stocked with books and had insufficient space for study. In many cases, classroom, laboratory and teaching equipment and furniture were inadequate and in a poor state of repair. Provision of farm livestock, workshop tools and farm machinery was insufficient. The report also observes that the faculties of agriculture have developed in the British tradition of scholarship, with considerable staff freedom and autonomy in carrying out teaching and research. This may be without reference to national needs. There is insistence on good academic performance by students, and Nigeria can call on a substantial pool of well-educated agriculturists, particularly for agricultural planning, administration and research. However, teaching programmes tended to be inflexible, and insufficiently up-to-date in both technology and method; key areas such as irrigation, agricultural teacher training, farming extension and farm management were neglected.

One must confess uneasiness at attempts by governments to direct and regulate teaching and research programmes at universities. Past experience gives little confidence that a government can select priorities better than anyone else. All institutions might be harmonised into the same rut! However, a government could allocate special research funds in fields thought to be important, and express forcible opinions on the achievements of university and institutional research. It must be remembered that, compared with research institutes, university research workers are often very inadequately supplied with supporting staff and facilities, and may be overburdened by teaching and other duties. Their research may need to reflect teaching needs, rather than farming needs.

The problems are not peculiar to Nigeria, and can be seen in other African countries, both anglophone and francophone. Scientific study and research are difficult without a good library, access to up-to-date journals, and to a meticulous and properly classified card index. (A wholly computerised index may be too risky.) Though efficient libraries remain, decay is evident in many institutes and universities. Books and journals are missing, or have pages removed, or are wrongly classified; they may be worm-eaten and are frequently misplaced on the shelves; there is a dearth of new volumes, and journals have been allowed to lapse. The card index lags well behind new accessions to the library, sometimes by several years. The staff are negligent in their duties, and a sight of the Head Librarian in the body of the library, trying to rectify defects himself, is rare indeed. Needless to say a good agricultural librarian should be interested in, and knowledgeable about, agriculture.

Agriculture is essentially a practical subject, and university farms and plantations are needed for practical demonstrations, for work practice, and for investigational studies and experiments. Yet, not infrequently, they have the appearance of bankrupt estates, with dilapidated buildings and fences, deteriorating roads, unkempt fields and graveyards of machinery. Abandoned piping from overhead irrigation may lie scattered, and overgrown by weeds. Screened plant houses stand empty, with torn netting and decaying benches. The reasons may be that responsibility for the farm has been entrusted to a person selected for his seniority, rather than for appropriate training and experience, who may be too busy with other matters; sometimes he is reluctant to become directly engaged himself in the field work. Other reasons lie partly in the general lack of funds, partly in the employment of too many staff and admission of too many students, therefore leaving insufficient money to maintain equipment or do any work. Even with the best will in the world, it may be impossible to keep sophisticated equipment operational – because there is no-one to mend it, or if there is, the parts are not available in the country, and because of import and exchange control regulations it is impossible to buy them overseas. Students may be unable to travel to see different techniques on actual farms and Research Institutions, because there is no money for transport.

It would certainly be an improvement if there were fewer students better taught. My impression, in Nigeria, is that the tremendous expansion in student numbers has been at the expense of teaching standards, and that a deterioration in the quality of graduates may be beginning to be noticeable. Stiffening the teaching staff with suitable expatriates might help – but even if acceptable would require a considerable technical aid input. Technical aid organisations themselves tend to be inefficient and expensive. A restriction of administrative, clerical and service staff to nearer ten per cent of total staffing, as recommended by Professor Mellanby (quoted in Chapter 6), would release funds for teaching and research, and might improve the overall efficiency as he suggests. As the number of ancillary staff increases, so does their influence within the institution. This may end with it being run for the sake of the bureaucracy rather than of the work that is supposed to be performed – and Europe has not been exempt from this. It may be better to put people to making roads, or planting trees, or breaking stones, as a sort of welfare relief, than to staff bureaucracies with them!

Simkins writes, 'There is another less discussed reason . . . This lies in the bureaucratic culture of government which ensures, not merely that policy-making and planning are centralised, but that the whole process of administration and management is organised in a way which tends to inhibit the kinds of development which are necessary if the quality of educational provision is to improve. This bureaucratic culture, which is not confined to Ministries of Education, places its . . . emphasis on hierarchical structures of decision-making, and on control through rules, regulations and routines. Such a culture may be quite efficient in maintaining patterns of educational provision which are well established, have clear purposes, and can be easily standardised . . . But it can be quite inhibiting where the major requirement is for creativity, innovation and responsiveness to diverse needs . . . There is a second implication . . . which is particularly significant for education. That is that status becomes attached to office rather than to ability, and the role of the professional tends to be seriously diminished. This expresses itself in various ways. One, of course, is the aspiration of so many teachers to become administrators. More debilitating, however, is the tendency for those responsible for educational provision . . . to see themselves as administrative functionaries rather than as professional educators. The role of school principal, for example, is defined simply as a junior position in an administrative hierarchy, concerned primarily with the routine implementation of rules promulgated by the Ministry, rather then as a professional concerned with developing the quality of provision in his or her institution, within a broad framework of policies and objectives established elsewhere. The bureaucratic culture, therefore, is essentially inhibiting of educational quality. It leads to the blind application of regulations rather than to the . . . sensible adaptation of policies to particular circumstances; to the referring of decisions upwards rather than to the taking of initiatives; and to evaluation of performance in terms of compliance with rules rather than in terms of achievement of quality in educational provision.'[12]

In universities, and in polytechnics and technical schools, in both anglophone and francophone Africa, each student studies a large

12 Simkins, T. (1985) *Managing education, West Africa,* No 3518, 28 January 1985 (pp. 158–160)

range of subjects, divided into a multiplicity of courses. Each course is separately examined at the end of each term or semester. In consequence, students spend too much time sitting examinations, and too little time on practical work or thinking. With amazing facility, they forget one course in order to learn the next! The separate courses are not integrated in the students' minds. They concern themselves only with passing examinations, and gathering the requisite number of course marks, and appear almost totally uninterested in genuine learning. Lectures are preferred that can be regurgitated in the examination room; teaching that tries to provoke thought is irritating or confusing. Though students spend a lot of time writing in the examination room, the papers are not returned to them after they are marked. Students receive little essay practice, and often have had a poor foundation in composition at their schools. They may voice the opinion that they are doing an applied science course, not studying the English language; but consequently the written work of many students is of poor quality. The process of writing clearly also helps clear thinking, and students sometimes get a course in 'Report writing', but this may be entirely theoretical. The result is graduates who cannot express themselves in writing, who have not learned to integrate their thoughts, and who cannot discipline their minds – perhaps not the sort of stuff that innovative farmers can be made from, and if nobody is worried, they ought to be.

Dr Johnson wrote, 'To the strongest and quickest mind it is far easier to learn than to invent. The principles of arithmetic and geometry may be comprehended by a close attention in a few days; yet who can flatter himself that the study of a long life would have enabled him to discover them, when he sees them yet unknown in so many nations, whom he cannot suppose less liberally endowed with natural reason, than the Grecians or Egyptians?'[13] Probably Johnson regarded England, in his day, as being inferior intellectually to ancient Egypt and Greece. With the fall of the Roman empire in the 4th century, the barbarian world virtually destroyed the civilised world – at least the cultural pool based on the Mediterranean – and Europe went 'back to bush' for more than a thousand years. The foundations of modern society were laid by the reawakening of scientific and intellectual thought in the 17th and (particularly) 18th centuries.

There are, of course, exceptional students who can survive almost any degree of mal-education, but complex systems of calculating

13 Johnson, Samuel (1951) Essay from *The Rambler* No 154, Saturday 7 September 1751

marks, averaged and weighted over all the courses, may reduce differences between students and make class position almost a lottery. It seems to be arranged so that few individuals fail, but real merit may scarcely emerge either. University is an elitist education; do we wish to grind everyone to a common denominator? Nevertheless it gives great pleasure to see, in Africa, young men and women dressed in much the same way as their kind throughout the world, enjoying something of the privilege that meant much to me when I was a young man. (And in England then, university education was still exceptional and elitist.) Despite the trials and tribulations they suffer, both in accommodation and working conditions, students still have that unique opportunity to study hard, play hard, and for a few years to enjoy to the full the stimulating company of other young people. Moreover, as university education has become an increasingly normal part of the African scene, and as world recession has worsened the economic situation, students have become less arrogant and even more eager to learn. They will be very disillusioned if they do not receive an education commensurate with their hopes.

Education and economic development

Lugard, writing on the desirability of establishing university education in a developing country, said, 'The financial support which the ... government can legitimately give, from revenues raised by general taxation, to a project which benefits a comparatively small section, must necessarily be limited by the imperative demands for popular education.'[14] We may note that Lugard was personally interested in promoting university education, and helped to found the University of Hong Kong, when he was Governor of the Colony, although he himself was an army officer and never attended university. Galbraith writes, 'We think of economic progress as the investment of present resources for increased future production – the investment of savings for growth ... Education is both a form of consumption and a kind of investment ... When we think of education as a consumer service, it becomes something on which we should save ... But when we think of education as an investment, it becomes something not to be economised but emphasised. The enrichment of the mind is as important as the nourishment of the body. Intellectual activity is

14 Lugard, F.D. (1922) *The Dual Mandate in British Africa*, Blackwood, reissued Cass, 1965 (p. 643)

properly pursued for its own sake ... In these attitudes, education, though a superior consumer good, is still a consumer good ... But there is another view ... outlays for education may bring large increases in production ... To rescue farmers and workers from illiteracy may certainly be a goal in itself. But it is also a first indispensable step to any form of agricultural progress. Nowhere ... is there an illiterate peasantry that is progressive. Nowhere is there a literate peasantry that is not. The notion of consumer sovereignty [i.e. the right of the consumer to choose between various forms of consumption] when brought to education, suggests that the student who has reached the age of discretion, more or less, has the right to study or not to study ... But if in fact, the student is the privileged object of the investment of scarce resources, the matter is not so clear. Society has given him some of its savings. Surely he has a clear obligation to return to society the increased production that society expects and for which it has spent its scarce substance? ... Colleges and universities must be responsive to the needs and requirements of economic and social advance and must be organised to make this possible.'

He continues, 'The schoolmaster, at his best and in the most democratic countries, has always been a rather authoritarian figure. In recent times, Latin American universities have been experimenting with highly democratic direction in which students, graduates and Faculty all participate more or less equally. It is a formula for deterioration into incoherence and chaos ... The new country cannot be so permissive [as the older and more developed countries] towards those in whom it invests. They are a privileged group who must work to deserve their privileges ... The developing country must consider its educational system in the light of the peculiar requirements of development. It cannot simply adapt from the older models. Having come late to development, it is the good fortune of the new countries that they can learn from others. But so much of what exists in other countries cannot be copied without serious cost. Adaptation is as demanding in its own way as innovation.'[15]

The intellectual elite of a country have a much more powerful say in the operations of government than the man in the street, or the peasant in the farm. They have a vested interest in university and secondary education, both for their own direct benefit (as possible employment) and for the benefit of their children (as education).

15 Galbraith, J.K. (1964) *Economic development,* Harvard University Press (p. 109)

However for the vast majority of people, primary education is all that they will receive, and it is their greatest need, both for their own benefit and for that of the nation. The cost of educating a primary school child is a fraction of that for a university student – particularly in engineering or the sciences. Primary education should become the birthright of every African, but university education cannot. Lugard, writing of former times, said, 'Provincial reports [in Sierra Leone] speak of the contempt for manual work shown by boys from schools. Education has brought to such men only discontent, suspicion of others, and bitterness . . . As citizens they are unfitted to hold posts of trust and responsibility where integrity and loyalty are essential, or to become leaders of their community in the path of progress.'[16] Today, primary and secondary education have been absorbed into African culture, at least in the more southerly parts of West Africa, and are not the disruptive factors they once were. However this is not yet entirely true of university education. To create large numbers of university graduates, without the prospect of employment commensurate with their expectations, is to produce unhappy and unfulfilled people, alienated from the society in which they find themselves. This is not beneficial either for the individual or for society. The primary school is the basis for education – and if this is deficient, prospects for further education may be blighted and the quality of secondary education will also suffer. It is in the primary schools that most of the citizens of the new Africa will be formed; primary education should not be the poor relation of higher education. Education does not end with leaving school, as perhaps Lugard himself demonstrated; and there may be many a millionaire to prove the point as well. It is not yet possible to give secondary and university education to all who deserve it, because of the cost, and there has to be elitism – either to those who can pay, or if at public expense, to those of exceptional merit. The numbers entering further education must be based on real (and not theoretical) manpower needs of the nation. Though this has been underestimated in the past, today the general supply does seem to be outrunning demand. During the past decade, the surfeit has been absorbed by continually expanding the size of the Civil Service. Now this has clogged the machinery of Government, and has added to the burden on the productive part of the economy – to which graduates ought to look for employment.

Adult education can expand literacy in the older generations, and

16 Lugard, F.D. (1922) *The Dual Mandate in British Africa,* Blackwood, reissued Cass, 1965 (p. 643)

help to advance the whole educational tone of a community. Once stimulated, the desire for intellectual improvement can become compulsive, and continue throughout life; this is to be encouraged. Primary school teachers could also be trained to give after-school teaching to adults; this would be cheap, besides providing a source of additional income to the teachers. However, in Jamaica, about forty per cent of school leavers are inadequately literate and numerate, and have to return to adult education for remedial instruction. It is definitely preferable that primary school leavers should have been taught properly in the first place. Adult education can improve the quality of life for older people, but it is the younger generation in particular that holds the key to the future.

Dr Dorothy Njeuma, Vice-minister of Education in Cameroon, said, 'In Government schools in Yaoundé and Douala, Cameroon's largest cities, although the teacher-pupil ratio is 1:42, the average number of pupils to a classroom is about 1:140. Most of the schools, therefore, run on a double shift system. Besides, a good 50 per cent of the teachers are unqualified. This factor, coupled with that of over-crowding, means that the efficiency of the system is rather low ... drop out and repeater rates are high – about 28 per cent on average – and the average length of time a child takes to complete the six or seven year primary school course is 10 years'. Yet between 1981 and 1984 Cameroon spent 18.2 per cent of its national budget on education.[17] Moreover, academic standards in francophone countries including Cameroon, are perhaps higher than in anglophone countries of West Africa generally.

Teachers must be restored to their former position of respect in society. This loss of status may have been because government itself has treated them shamefully, for example by failing to pay them for months together at times of mismanagement of funds, and perhaps also because of the behaviour of many of the teachers. Teachers become distracted by other means of earning a living, such as trading, taxi-running or chicken-farming. But how can pupils be expected to develop a sense of responsibility, when teachers themselves take a neglectful attitude to their duties? Is this also because school governors and school principals have lost control of staff appointments, promotions and discipline, when schools were taken over by the Ministry of Education? Is it because the moral

17 Njeuma, Dorothy (1985) in a paper delivered at the conference on *Education priorities in Sub-Saharan Africa,* Windsor, England, quoted in *West Africa* No 3549 of 2 September 1985

influence of the religious organisations is removed, and teachers are posted hither and thither like civil servants? Though the motives of achieving secular education, and the difficulties that rural schools have in obtaining staff, are appreciated, one wonders if the cure has not been worse than the disease. There are, of course, many honourable exceptions, and I confess that I personally would not find school teaching an easy career; but it is school teachers particularly, as well as university teachers, who are helping to lay the foundations of Africa's new society.

Russell wrote, 'From early childhood parents and friends are hypnotising the child into beliefs and unbeliefs, and too often they are exiling all nobility out of life, all confidence, all trust, all hope; they are insinuating a mean self-seeking, a self-mistrust, a vulgar spirit which laughs at every high ideal, until at last the hypnotised child is blinded to the presence of any beauty or nobility in life. No country can ever hope to rise beyond a vulgar mediocrity where there is not unbounded confidence in what its humanity can do.' He also wrote, 'Mr Edmond Holmes tells us he found, when he was inspector of national schools in England, a backward country district where the genius of a teacher divined a soul and a kinship with immortal things in the children ... She bent herself to liberate these powers, and a crowd of lovely things went fluttering out of the opened cage. There the feet of the little rustics danced as the feet of life should dance. They loved and saw beauty ... There art and music and literature were loved. There was imagination, happiness, and quick intelligence ... Yet all this vision and beauty did not make the children unfit for labour afterwards, for on enquiry among the farmers it was found that no sluggards or lazy workers came out of that school. The evocation of the higher faculties in men or women does not unfit them for the world's labour.'[18]

Endnote

'In Sierra Leone in 1978, enrolment in primary schools was 227 800 of which 39 per cent were girls; and in secondary schools was 54 000 of which 32 per cent were girls, secondary school enrolment having increased 8 times in twenty years. At university in 1980 there were 1600 students of which 17 per cent were women, and in 1983 this had

18 Russell, G.W. (1912) *Co-operation and nationality – a guide for rural reformers from this to the next generation*, Maunsel & Co, Dublin, reissued by the Irish Academic Press in 1982

increased to 2400 students of which 20 per cent were women, for a total population of 3½ millions.'

West Africa No 3498 of 3 September 1984

'In Ghana, during 1983, out of a total national population of 12½ millions, the annual enrolments were 200 000 into primary school, 25 000 into secondary school and 2400 into university. The average annual cost of educating a primary school pupil was estimated as ¢510; but ¢2500 for a secondary school student and ¢53 000 for a university student. About 15 per cent of total educational expenditure was spent on universities.'

West Africa No 3508 of 12 November 1984

In Nigeria the oil boom has encouraged a "democratisation" of education. For a population of about 95 millions, there are 15 millions in primary school, 5 millions in secondary school and almost 100 000 at university. In twenty years, numbers at primary school have increased 5 times, at secondary school 20 times, and at university 30 times.

J. Osuntokun
Expansion of university education in Nigeria, problems and prospects
National Conference on Nigeria since Independence
Ahmadu Bello University, Zaria, 1982

8 The implications for society

The foolhardy, the unlucky and the impoverished will become landless. It is inevitable that as development proceeds there will be a progressive separation of the greater proportion of the population from the land. Land will become too valuable a resource to use as a dole for the otherwise indigent members of the community.

'It has long been the fashion to speak of the African as naturally lazy, leaving work to his women, and contented to lie in the sun and eat and drink. It would seem, however, . . . that there are few races more naturally industrious. It is the tradition of Africa that the men tend the cattle and fight, while the women work in the fields, but throughout the greater part of the continent the men share the fieldwork with women, and in many districts do by far the larger part of it.' Lugard then quoted Captain C.J. Orr, 'When the African . . . is given an incentive to work, he will work in a way that is sometimes almost astounding . . . Give him an interest in his task, encourage his initiative by making him think for himself, thrust responsibility upon him, demand results, and not the mere mechanical performance of labour, and he will be found surprisingly industrious.' Lugard also wrote, 'The aptitude of the African as a skilled worker is abundantly in evidence in the railway, printing, and other workshops. In West Africa . . . he has replaced Europeans in many departments of skilled work, and he does not require an undue amount of . . . supervision. Even in highly technical work such as acetylene welding and cutting, in pattern making from dimensioned drawings, and in the use of the lathe and of modern machine tools, he bears witness to his efficiency.'[1] Writing in the early 1920s, for the benefit of his own compatriots, Lugard had no need to be mealy-mouthed, or to say

1 Lugard, F.D. (1922) *The Dual Mandate in British Africa,* Blackwood, reissued Cass, 1965 (p. 643)

what he did not believe. Yet today, in offices throughout Africa typists are sitting staring out of the window, or slumped with their heads resting on their machines. The boss is reading the newspaper, or (if it is functioning) talking on the telephone to friends; or he has 'gone to the bank', or is 'not on seat' and nobody knows when he will be back. The official vehicle is broken down, or without petrol, or is being used for frivolous duties in the town. The labourers are sitting in the shed talking, or cultivating the boss's personal farm; or they have dispersed because they have not been paid. The tractor is waiting for a spare part, but no company will honour the purchase order, because of the difficulties in obtaining payment, and nobody can afford the cost of advancing the money; or the part is nowhere available to be bought. Is this exaggerated? Perhaps, but it is not unrepresentative. What has gone wrong?

Do Europeans do any better? They certainly may have the knack of *looking* busy, and North Americans, as a British Army sergeant once said to me, 'Buzz around like blue-arsed flies'. They will struggle manfully, though not necessarily successfully, with a situation that most Africans have already given up as a bad job. *WAWA* means 'fool' in Hausa, or 'West Africa Wins Again'! Of course they may seem to succeed for a while; but Elspeth Huxley wrote, 'Africa is like a crocodile that swallows people, cultures, ideas, and then breaks down, digests, disposes, builds their substance into its own; and in the end there is just the crocodile, with its cold unwinking eye.'[2]

Africa does change. Though in many ways there is still much that is the same, it has altered a great deal since I first stepped on its shores thirty years ago. African society is often regarded as bound by tradition and slow to change; in fact it has been astonishingly adaptive without losing cohesion. In former French territories the elite may worship *francophonie,* but yet the basic African identity flows strongly in the national character. It will emerge even more strongly as the French tie loosens. In France 'the State' is supreme, whereas for the Englishman the individual is sacrosanct. The British did not feel the same mission to tamper with the culture of their colonial subjects, except in matters which they considered morally objectionable, such as domestic slavery or the killing of twins. Of course the association with Britain, tenuous though it was for most Africans, has to some extent modified cultural attitudes and behaviour. Dumont wrote, 'British territories were dominated by an outdoor-loving non-

2 Huxley, E. (1954) *Four Guineas: A journey through West Africa,* Chatto & Windus (p. 303)

intellectual public school Oxbridge-type, conscientious (apart from the Kenyan pro-settler attitude) in protecting their wards; the French were dominated by literary Sorbonne products longing to get back to Paris and thus rather less willing to incur the displeasure of metropolitan vested interests.'[3]

It must also be remembered that, whatever his racial differences from the European, the African has always belonged to the ancient cultural pool that centred on the Mediterranean. The north European and the tropical African were at the extreme poles of the same culture; that is, until Europeans short-circuited it by travelling round by the ocean. In African society there is usually an excellent information flow, both upwards and downwards, and this is the essence of democracy. It is much greater than, for example, in England, though it is inevitably burdened by inaccurate speculation and rumour. In 1879, Rider Haggard was amazed when, thirty hours after the event, his elderly African washerwoman told him of the annihilation of a British Army column by Zulus, 200 miles away at the Battle of Blood River (i.e. Tugela River) near Rourke's Drift. The Europeans were ignorant of the event until some of the few survivors reached them three days later.[4] Possibly he thought she was clairvoyant, but my guess is that the news travelled via the talking drum, which imitates human speech and can relay information over long distances. Perhaps, today, it has been replaced by the transistor radio. Ballot boxes are only recent trappings of democracy. I have experienced a military government that was, at times, more democratic than an elected government had ever been.

In Africa, to travel back a century in time would be for most Africans to journey into another world, though a European doing the same in his own country might find much that was similar. Peasant societies depended for survival on cautiousness and on mutual responsibility, but at the same time they had to be adaptable enough to accommodate change. In Africa, customary law has been highly sensitive to changes in social mores. The reasons are that the law is neither written, nor codified, nor operated from precedents – though precedents are respected to some extent. The remedies for many of Africa's ills lie in socio-economic reforms. People are capable of taking these measures for themselves, as they did formerly in England, and are trying to do again in Nigeria. Governments may hinder

3 Dumont, Rene (1966) *False start in Africa,* Andre Deutsch

4 Haggard, Lilias (1951) *The cloak that I left, A biography of the author Henry Rider Haggard KBE,* Hodder & Stoughton (p. 288)

progress by trying to hold on to the past. In fact, governments may have more power to do harm than they have to do good. Allow people, as far as possible, to work out their own salvations, and millions of minds will be turned to the task; not simply the brains of the few men who head government. The adaptation of Africans to changing circumstances has revealed a high degree of intiative and entrepreneurism latent in African society. We must restore to modern African society at least the adaptability that traditional society formerly had.

Advances in scientific knowledge are releasing African agriculture from the biological strait-jacket in which it has been constrained. Admittedly the problems are more difficult than those which impeded farming in Europe at the same stage in its development; North America has never had to face the same sorts of problems at all. There is need to mobilise farmers to take advantage of this information in order to increase productivity so that Africa can at least feed itself and supply home industries with raw materials, and to earn the foreign exchange to purchase those things that Africa is unable to provide for itself. Allan writes that the role of agriculture in the early stages of development is fourfold:

- to generate capital for the economic transition, including foreign exchange gained by the export of soil products;
- to provide the labour force required for expansion of the non-agricultural sectors of the economy;
- to meet the growth in demand for agricultural products created by industrial consumption, an increasing non-rural population, and rising per capita incomes;
- to support a surplus population until such time as this can be absorbed by the general expansion of the economy.[5]

The degeneration of African agriculture

Peasant farming conserved the soil, but is breaking down because traditional farming is insufficiently remunerative to satisfy modern aspirations for higher living standards. As discussed in earlier chapters, young people want to earn wages in order to have money in their pockets, and to escape the bonds of peasant society. Growing populations have been fed by increasing the area cultivated, but this leads to a shortening of the fallow period and hence less time for soil

5 Allan, William (1965) *The African husbandman,* Oliver & Boyd (p. 505)

recovery and a reduction in the area of 'bush' for pasturing animals. Government is trying to counteract the resulting loss of fertility by encouraging farmers to use artificial fertilisers, and to extend the area cultivated by using mechanised cultivation; but these measures do not restore the humus which, in tropical soils generally, is responsible for the ability of the topsoil to retain nutrients and structure, against the leaching and erosive effects of rain and wind. In more humid areas a consequence is sheet erosion and gulleying, and in arid regions the topsoil is being blown away. The problems have been exacerbated by the removal of the protective cover of natural vegetation, and by heavy machinery compacting the soil, causing waterlogging and poor drainage, and interfering with root penetration. Modern agricultural chemicals are introduced, including insecticides, fungicides and selective herbicides. The sprays are dangerous, and protective clothing is uncomfortable in a hot climate. People have become ill and have also died through using the empty containers for domestic purposes, disposing of them carelessly and contaminating water supplies, or using insecticides as fish poisons or to preserve food.

Attempts are also made to help the farmer with storage by building communal silos for grain, which usually seem to be disused, or to help with marketing by establishing a grain marketing board which benefits the consumer, or its own bureaucracy, more than the farmer. Hyper-expensive irrigation projects are begun, which require relocation and regimentation of human populations by bureaucrats, leading in at least one instance to serious rioting and bloodshed. Such schemes are often dependent on teams of foreign experts, and armies of extension agents, advising illiterate or semi-literate farmers, who will still earn a wretched income, even if less wretched than before. When the foreign experts go away, when the extension agents stop touring, when the fertilisers and agricultural chemicals are not available, when the tractors and pumps break down, the farmers sink back to their former ways – well – almost to their former ways, for there are always some lasting changes and improvements, though these are rarely commensurate with the cost. What are the reasons for failure?

Problems of industrial methods

Perhaps the lack of success is because of the attempt to introduce some of the methods of industrial farming, used in Europe and North America, into what is essentially still a peasant society. These industrial methods of farming have been developed within industrial societies; but what may seem so effortless in an industrial society is nothing of the sort in a peasant society. To the foreign expert it may all seem simple enough, and so in a way it may be, but it just doesn't seem to work. Industrial societies became industrialised before they introduced these methods into their agriculture. Africa is not industrialised. Peasants do not have the same attitudes to promptness, to precision engineering, to organisation and discipline of their society, as exist in industrial society. Peasant societies *are* disciplined and organised, but not in the same sort of way. They may or may not possess the technical skills, but the difficulties are more subtle than this, and reside more in human relationships. In a peasant society one could say that human beings are more important than the efficiency with which the work is carried out. In an industrial society it is the other way round: the work takes precedence over human relationships. In an industrial society, social relationships are largely practised 'after hours', and may be substantially restricted to close relatives and neighbours. At work, social formalities may be perfunctory; Americans dispense a veneer of bonhomie, but can be much more brutal in their dealings with one another – *hire and fire* – than is usual in Europe, where there remains a residuum of earlier communal habits. People are not obliged to be responsible for the livelihood of the other members of their community. In many ways though, people, and particularly young people, prefer the more impersonal relationships of urban society, and the relative freedom that it gives to the individual.

Russell wrote, 'The factory system was invented and was an instantaneous success. Men were clothed with cheapness and uniformity... They were at their desks at nine o'clock, or at their looms at six. They adjusted themselves to the punctual wheels. The rapid piston acted as a pacemaker, and in England, which started first on the modern race for wealth, it was an enormous advantage to have tireless machines of superhuman activity to make the pace, and nerve men, women and children to the fullest activity possible. Business methods had a long start in England, and irregularity and want of uniformity became after a while such exceptions that they were regarded as deadly sins. The grocer whose supplies of butter did not arrive week after week by the same train, at the same hour, and of the same quality, the same colour, the same saltiness, and in the same kind of box, quarrelled with the wholesaler, who in his turn quarrelled with the producer. Only the most machine-like race could win custom ... In Ireland the wild child humanity came under this discipline very slowly ... He skimmed his milk himself, or his wife did, and he or she pounded away at the churn and were unconcerned about temperature, and they had never heard of microbes. There was nothing uniform about the produce, and in a hundred farms in every parish a hundred flavours originated. They might be all bad, or all good, but they certainly were never the same. Yet the Irish farmer was manufacturing it for people who had standardised all their tastes and objected to the slightest change, who wanted all their butter to be as uniform as all the square yards in a roll of wallpaper. The exceptional excellence of the butter received one week was no compensation for the intolerable flavour of the next. Denmark, twenty years before we began, turned itself into a machine, and its farmers all got up at the same time in the morning and took their milk to creameries, and cream was separated and churned at the same temperatures. The managers had heard all about bacteria ... and they kept their premises as clean as the wards of a hospital, and they turned out butter so uniform that all the Danish butter might almost have come from the one churning. The individualist in butter-making was nowhere.'[6]

Peasants had disappeared from England before 1900, though they continued to exist in many West European countries until after the Second World War. This was partly responsible for the early success of the European Economic Community, which was able to recruit these people into expanding industrial development. In Britain this was no longer possible. In the early 1960s I remember a colleague importing a *Ford Taunus* car that had been manufactured at a new industrial plant in Germany. It was found to have three leaves in one rear spring and four in the other; the rear window fell out of its frame, and the rest of the car gradually shook to pieces. In those days, workers at the British Ford Factory at Dagenham could turn out better cars in their sleep! But twenty years later this is no longer true: the German Ford is at least as good, if not better, than the British model. The point I am trying to make is that, in industrial communities, there exists a sense of organisation, of discipline and of duty, in addition to technical ability, that enables them to function smoothly; and these are a part of what I have called the *industrial ethic*. Peasant societies have to learn these qualities the hard way. Some lucky countries, perhaps the more militaristic ones, may have them already to a useful degree within their peasant society. Nevertheless it is probable that many West European countries were formerly no better endowed in this respect than African countries are today. They were afflicted by

6 Russell, G.W. (1912) *Co-operation and nationality – a guide for rural reformers from this to the next generation,* Maunsel & Co, Dublin, reissued by the Irish Academic Press in 1982

the same ramshackle attitudes to organisation and to machinery, and to ensuring continuous electrical power, water and telephones. At present, it may verge on the impossible for Africans to maintain industrial systems in Africa. An individual may have all the attributes necessary for success, but unless he is supported by sufficient other individuals with the same abilities, he is like a man trying to bale out a badly leaking canoe – he plugs one leak only to reveal another, and then another and another . . .

The need for mixed farming

The solution of the problem is to be like the peasant farmer, and *not* to depend on other people for the success of one's efforts. Grow fodder crops, keep more livestock, make more manure to fertilise the fields, and grow more grain. Use draught animals for cartage, for ploughing and harrowing – at least to supplement the tractor, and to replace it when necessary. Rely on rotation of crops to control weeds and diseases, and supplement this with agricultural chemicals if they are available. Burn oil lamps if necessary, use hand tools. This is the way most farms were in Europe during my childhood. There were no telephones, no piped water (unless diverted from a convenient stream), no electricity, not always a tractor, and horses were still responsible for much of the traction. I am not recommending it, but simply reminding people that it is possible! In Africa the remedies may not be exactly the same as in England, but the logic is the same, as the chapter on farming has attempted to explain.

More recently, 'mixed farming' is being reintroduced into America and Europe as 'organic farming': fertilising the soil mainly by means of organic manures, which help to maintain the humus content of the topsoil. As earlier chapters have tried to show, this is also an answer to the more difficult biological problems of farming in the tropics. The changes in farming that are needed do not necessarily require much in the way of additional skills for the farmer, but rather a reformation in his conception of agriculture.

1 Agriculture is required to supply greater quantities of meat and animal products, as well as of staple foodstuffs, to burgeoning urban populations – rather than to provide subsistence to isolated rural communities.

2 Formerly 'bushmeat' was an important part of the diets of African peoples; but the increase in human populations minishes the area of natural habitat available to wildlife; and

more hunters and better communications speed the massacre of wild animals.

3 Instead of increasing the cultivated area, at the expense of rangeland for livestock and for game, and hence reducing the number of animals that can be maintained, fodder crops are grown and the numbers of livestock may actually be increased.

4 The essential feature is to make the fallow productive, in order to augment the yields from an existing area of farmland.

5 Livestock are more valuable, per unit area of land, than staple food crops, and therefore give greater possibility for profit from a given area.

6 The greater production of manure from more numerous livestock can be used to improve soil fertility, and hence to boost the yields of both food and fodder crops.

7 In the humid tropics the natural fallow may be trees, and it is possible to incorporate trees in the cropping systems, both as a source of wood, and sometimes as fodder.

8 Mixed farming has the advantage of distributing the work load over the year, and of spreading the economic and natural risks.

These factors assist in making mixed farming attractive.

The need for industrialisation and the role of apprenticeship

In earlier paragraphs it may have appeared that industrialisation in Africa was being deprecated, but this is not the case. Progress in agriculture cannot be separated from progress in industrialisation. Of course it is easier to implant an industry into an industrial community, and to get it to function efficiently, than into a peasant community, even within the same country. This is well known and amply demonstrated by experience. It is often necessary to attract industries to non-industrial regions by means of cheap labour, cheap land, and sometimes also by tax concessions and grants. Children born in a town have an advantage as future industrial workers. From an early age they have been able to fiddle with worn out clocks, broken bicycles, junked motor cars, and all the debris of modern society. In developed countries, parents and schools go to a great deal of trouble to develop the manual and visual skills of their children, by the games and pastimes that they encourage them to play – jigsaw puzzles, Meccano construction sets, model railways, toy motor cars, personal bicycles, drawing and painting. They are given books to read, and have access to a public library. Today this luxury of paraphernalia even extends to walkie-talkies and computers. Meanwhile the peasant child in a rural village has not much beyond corn stalks, sticks and stones from which to make his toys.

I remember having the task of building a pre-fabricated house with the help of a messenger (a former army sergeant) and some peasant farmers. No construction plan was provided, nor explanatory diagrams, and we had never seen a completed house. There was a page of type-written instructions, of which a sample read, 'Fix a truss and a pair of knee braces at the junction of the assembly frames. The tie-bar of the truss fits to the upper bolts in the ribs of the roof panels. From the centre of this tie-bar struts rise to each end of the cross member and are bolted to the underside of it'. We were provided with a miscellany of panels, and various lengths of angle rods. Trusses and knee braces might have been surgical appliances for all we knew. All the pieces were laid out on the ground, and I stared at them and considered how all might be fitted together. Eventually I realised that there were the pieces of two buildings, including a separate kitchen. The farmers could never have succeeded in solving this puzzle. I found that I could screw four nuts and bolts together in the time that each of them could screw one. They seemed to have been equipped with a handful of thumbs instead of fingers! But I also remember my son, who had been born and brought up in Africa, being sent back to school in England at the age of eleven. He was taken to buy warm clothing for the winter, and was given gloves to try on for size; but he was simply unable to get his thumbs into the thumb holes and his fingers into the finger holes, at least not singly! The shop assistant gazed at him in horror, evidently concluding that he was mentally handicapped! In embarrassment, my wife hastily paid for the gloves, and took him home where he could try them on in privacy. He has since learned to be a firstclass cabinet worker, and is presently studying engineering at a famous British university.

In Africa, there have long been the traditional skills of blacksmithing, bronze casting, goldsmithing, woodcarving, basketry, mat making, weaving, gown making, embroidery, leatherwork and pottery. To these have been added carpentry, metalworking, masonry, printing, photography, motor repair, welding and soldering, painting and decorating; and these are nowadays widespread. They have been fostered by a customary system of apprenticeship; though the lads are selected more by family relationships, or by a parental desire that they should learn, rather than necessarily by innate ability. This has readily fused with the more formal British system of apprenticeship, to become an extremely important method of promoting and disseminating technical skills in anglophone Africa – which seems to be lacking in francophone countries. The master (or mistress, in a sewing shop) is encouraged to take pupils because, in exchange for being fed and housed, they provide a source of cheap bonded labour. Indeed they may have to pay for the privilege.[7] Many intermediate vocations, particularly manual, are included such as typing, acetylene and electric welding, carpentry, motor repair and so on. It is realised that the more modern skills are often badly taught by unqualified instructors in extremely inadequate conditions. Nevertheless the system has made an enormous contribution to African social and economic development, and this must be recognised. The skill and dexterity shown by some of these artisans is quite remarkable, and would be unusual even in Europe. African governments would do their countries a grave disservice if they try to sweep it away on the pretext of modernity and progress. Rather, they should attempt to encourage, improve and extend the existing practice. This could be achieved by strengthening the official trade testing systems, for which the clients could and would pay. One of the banes of education, and not in Africa alone, is that it is dominated by government (either national or local) and becomes very vulnerable to the activities of intellectuals and theoreticians, who may be more motivated ideologically or politically, than from practical experience.[8]

The need for expatriates

The robustness and self-dependence needed in agriculture have also to be fostered in industry. This is why plantation agriculture has long been able to function efficiently in underdeveloped countries, producing palm oil, crumb rubber, tea, tobacco etc. Agricultural and industrial production are directly integrated together. Clearly, industries based on raw materials produced within a country are less vulnerable than industries dependent on imported raw materials. The

7 In England the system included 'farm pupils', who were sent to study under, and work for, a practising farmer.
8 A quotation attributed to Gaius Petronius (AD 66) reads, 'We trained hard – but it seemed that every time we were beginning to form up into teams we would be reorganised. I was to learn later in life that we tend to meet any new situation by reorganising, and a wonderful method it can be for creating the illusion of progress while producing confusion, inefficiency and demoralisation.' It is always preferable to effect changes in ways that allow a strategic retreat, in case they are found to be mistaken. To lift both feet off the ground at the same time is to risk a tumble!

first industries introduced in Nigeria were sawmilling, soap making, cotton ginning, spinning and weaving, and rubber tyre remoulds. More recently, the emphasis has shifted towards the manufacture of consumer goods heavily dependent on imported raw materials: bicycle and motor vehicle assembly, radio assembly, brewing, biscuit manufacture, stationery and printing, plastic moulding, pharmaceuticals etc. It may be that this dependence of industry on importation is undesirable; but nevertheless these are the goods that people want, and at least part of the manufacturing cost is being met within the country. Moreover it helps to generate among the population the beginnings of the 'industrial ethic', and to lay down the necessary infrastructure. It may be dependent on borrowing expatriate managers and engineers from industrial countries to set up the new industries. Countries in tropical Africa do not feel able to encourage immigration of skilled people from more industrialised countries, as did America and Australia, who encouraged immigrants to settle, so that their children and the wealth they earned were absorbed into the new nation. In Africa, this could run the risk of creating an explosive situation of a predominantly white elite and a black proletariat. Meanwhile, in Europe, the need to protect Welfare States from wholesale immigration by refugees from impoverished Third World economies also compels official racialist policies. We must hope that time, by removing the causes, will allow the evolution of a more racially tolerant world.

Expatriates may also be needed to assist in developing profitable systems of agriculture, but the necessity to employ expatriates adds to the costs both of agriculture and of industry, which will need protecting from more efficient operations overseas. African countries are producing food primarily for their own domestic markets, at least in the first instance, and unlike Britain for example, they can afford to protect their own emerging industries. The protection is needed both for home-produced raw materials and for manufactured goods. Nevertheless there are ramifying effects through the economy, particularly if the protection is sufficient to affect the general cost of living, and to have adverse effects on the price of exports – unless of course export goods receive a bounty from Government. Dumont comments that an important characteristic of underdevelopment is the dominance of commercial capital over industrial capital.[9] The need therefore is to foster the transfer of wealth from commerce into agriculture and manufacturing industry.

9 Dumont, Rene (1966) *False start in Africa,* Andre Deutsch

Green revolution versus agrarian reform

The use of the term *Green revolution* in preference to *Agrarian revolution* is perhaps an attempt to dissociate progress in agriculture from the less palatable changes that go with agrarian reform – the concomitant needs for transformation of land tenure and for social reorganisation; but there is no alternative. If governments try to prevent the changes taking place, they will witness a progressive impoverishment of their populations. In tropical Africa, every person had a traditional right of access to land to farm and to build a house, in order to feed and shelter his family. This right has now to be abrogated to allow the development of industrial communities that are independent of the soil, and therefore have relinquished their say in the management of land. Future development of agriculture requires:

- freeing land of communal rights;
- consolidating land ownership;
- educating farmers;
- investing capital in farming.

Assuming the intention is not to nationalise the farming industry, and that farming is to remain private enterprise, then it is a minor progression from present systems of peasant proprietorship and inheritance of land to freehold ownership. Indeed, the universal evidence is that owner-cultivation is the most efficient form of land management. Another reason for its success is that it has facilitated capital investment in land, by giving the possibility of mortgage. This depends on rights in land being freely negotiable and alienable, and on having methods for conveyancing land which are quick, easy and cheap. This encourages the diversion of finance from commerce and industry into agriculture, and into the development of rural areas. Of course the money does not necessarily have to be used for investment in farming; the money raised by mortgaging land may also be used to foster other rural industries. In effect it makes possible the transfer of wealth from urban to rural areas.

There is a general abhorrance of speculation – not only in Africa – because of the feeling that it is objectionable for a person to reap where he has not sown. However, as explained more fully in Chapter 2, some speculation in land is necessary in order to create a market in land, so that it becomes potentially saleable and therefore mortgageable. Speculative hoarding of land should be prevented; this can be achieved fiscally by imposing taxes on land. These compel people to use land, or at least to pay for the privilege of not using it. The concept of 'freehold' is already well established in urban areas. It has the advantage that it:

- permits capital investment in buildings;
- provides a facility for carrying on a business, or for supplying a residence;
- allows mortgaging to gear up the growth of a commercial enterprise.

The idea has spread to rural areas in order to allow investment in agricultural enterprise. These developments occur quite naturally, and should be encouraged and speeded up by legislation to enable the 'enclosure' of land and its conversion to freehold, and by applying fiscal pressures. Nevertheless the change should be voluntary, undertaken on a local basis as and when a particular community undertakes to change. The timing will depend on the rate of economic and social progress within that locality. Farms can be either too big or too small, depending on the crops grown and on local circumstances. Thus the form of land tenure must be flexible, so that it may adapt to local circumstances, and so that farm size can respond to economic forces. There should be a view to arriving eventually at a common nationwide system of land tenure, by evolution from the varieties that exist at present in the different parts of the country.

Reinforcing the changes in land tenure should be the imposition of a national system of land inheritance, such that:

- male and female children inherit equally from their parents;
- husband and wife may own land separately or jointly, depending on agreement between them;
- inherited land is *always* divided equally among the children, excepting where it is less than a certain area (say 10 hectares), when it may be left to one child who must compensate the others (or some children may agree to relinquish their share and receive financial compensation); otherwise the land must be sold and the money divided equally among the children;
- a widow receives the proportion of land in trust for her children, which she may retain for her own use until her death, or divide equally among her children, at her pleasure.

Families may continue to operate traditional systems, but if an individual appeals to the courts he would automatically receive his rights under the new law. The intention is both to deter the formation of large estates, and at the same time to discourage the excessive fragmentation of land holdings. The aim is to promote owner-cultivation on holdings sufficient in size to make remunerative farming possible.

The foolhardy, the unlucky and the impoverished will become landless. It is inevitable that as development proceeds there will be a progressive separation of the greater proportion of the population from the land. Land will become too valuable a resource to use as a dole for the otherwise indigent members of the community, which perhaps is how it tends to be regarded at present. Other provision will have to be made for the destitute. In predominantly agricultural countries they must be maintained by means of a tax on agricultural production. It is better if this is based on the productive capacity of the land, rather than on its actual production. This compels people to use the land properly, or to lose it. Just as the land holding group was formerly responsible for providing land for all its members, after the metamorphosis in land tenure it becomes responsible for maintaining those who otherwise have no livelihood. Parochially administered funds are more likely to be disbursed honestly and equitably. From the able-bodied, the community may expect work in return for the subsistence it provides. It may be necessary for central government to set up a system of magistrates, backed by the force of law, to whom people may turn for prompt justice if they are treated unfairly. In the event of malpractice, the magistrates may be empowered by a special Government order to take over administration of the fund. Only those born within the community, or accepted either explicitly or implicitly as members of the community, should be able to come to it for relief. Compare the former parochial administration of the poor laws in England. Just as agriculture and industry must be robust and self-dependent, so it should be for poor relief in the rural areas. The tax on land should be graduated, so that it:

- bears proportionately more heavily on the larger land-owning bodies;
- encourages the breaking up of estates;
- promotes privatisation of land holding, by encouraging sub-division of ownership among individual proprietors;
- discourages speculative hoarding of land;
- fosters the small landowner.

Furthermore, the payment of tax over a sufficient period of time may be regarded as a proof of ownership, subject only to challenge in the courts.

The need for poor relief and the role of local government

As the proportion of the Gross Domestic Product accounted for by industry outgrows that of agriculture, then the support for the indigent section of the community has to be shifted from agriculture to industry. This is the basis for a *Welfare State*. All salaried and wage-earning employees are obliged to make compulsory payments into a provident fund for social security. Such funds already operate in francophone countries. At the initiation of such a scheme, the contributions from both employee and employer could be five per cent of salary. As the system establishes itself, the employer's contribution could rise progressively to twenty per cent of salary, and that of the employee to between ten and fifteen per cent. The purpose is to provide both unemployment pay and pension, though a certain period of contributions would be necessary before an individual could draw on the fund. Otherwise he would 'go on the rates' in whatever community he was born, or had been accepted subsequently. Rates would, of course, be payable in towns, as well as in rural areas; the cost of those falling through the social security net would have to be met from the rates. The compulsory savings generated by the fund provide an important source of capital for economic development. Where there may be difficulties administering the fund efficiently, management could be invited from reputable overseas companies.[10]

Though the national provident fund is intended primarily for those employed in urban areas, the same provision could be made for salary or wage earners in rural areas. Government ought also to apply a policy for a minimum wage, based on the cost of subsistence for a man, his wife and (say) three children, and obligatory for all men and women above a certain age – say eighteen years. Nevertheless, social security cannot cater easily for the self-employed and for peasants. Of course they could pay themselves a wage, and thus become eligible for provident fund contributions, but it is difficult to cater for peasants because they do not receive a regular income. They need a continuing system of relief based on parochially assessed land tax or rates. Moreover contributors have to earn enough, over and above their essential needs, in order to be able to contribute at all!

If relief is to be organised at the parochial level, then it will be necessary to strengthen local government. Unfortunately, over much of Africa, local government is in a sorry condition. As central

10 In Nigeria, a National Provident Fund was inaugurated in 1962 for all wage earners. However, in the mid 1970s, Government converted all its personnel to salaried terms of employment, and the operations of the fund became more or less confined to the employees of private companies. The fund tends to be treated as government revenue, rather than being directed into the private sector as investment capital, where it might be more beneficial for national development. Otherwise it becomes a tax on the productive part of the economy.

government has usurped more and more of the authority from the traditional system, and as the chiefs have lost their power, they have at the same time lost the respect of the people. Chieftaincy has become something that may be bought and sold. Meanwhile people turn elsewhere for guidance and advice, for the resolution of difficulties in human relations, and for the settlement of disputes; for example to the pastor of the local church community. The religious organisations are especially important in the towns, because of the community support they give to individual members, which may cut across tribal affiliations. Chieftaincy can buy lip service, but not true respect. It could be said that the very adaptability of the traditional system has allowed it to adapt itself out of existence! Nevertheless the problems must be faced, and local government must somehow be resurrected. In Britain, it seems that the system in which the electorate for local government is the same as that for central government, has undesirable consequences. National party politics are brought into the local arena, where they are not necessarily relevant to local needs. The irrelevance of the result is perhaps reflected in the lack of interest that people in Britain show in local government elections – by failing to go along to vote, unless sometimes as a sort of protest against the policies of the National Government. Should local government have an electorate different from that voting at national elections? Should it be elected at all? Perhaps a scrutiny of traditional systems and their methods may provide an answer. Traditional government represented the more mature and more responsible section of the community, which nevertheless tried to arrive at a consensus for the various pressure groups within the community, which included the various families and guilds, the women's and other societies, and the age group organisations. It could be required that both councillors and local government electors possess a property qualification, as formerly they needed in Britain, because local government has much to do with management of the property belonging to the community. In other words, those who provide the money could decide how it should be spent. Local government would still be subordinate to national government, elected by universal suffrage, and if it abused its functions could still be subjected to national legislation and to being over-ruled in the courts.

In anglophone Africa, a problem for local government has been that it is regarded by central government (and in Nigeria by State governments) as an extension of its own authority. This has resulted from the former colonial policy of 'indirect rule' – that is the central government using the traditional chiefs as agents of its authority. Government attempted to ensure that chiefs favourable to its policies were retained, and made them subservient to itself by paying them stipends, and (in effect) by discouraging traditional forms of remuneration. The chiefs were often made agents for the collection of government taxes, so that their people became reluctant to pay them twice: both the Government tax and the traditional exactions, and the latter fell into desuetude. Admittedly the government policy may have been to supersede undesirable customs, such as 'office' farms worked by slaves, but it did augment the dependence of the chiefs on the central administration. Nevertheless the lurking power of Government reinforced the position of the chiefs. A consequence was that, under colonial government, the power of the chiefs tended to become more autocratic than it had been formerly.

The advent of national governments elected by universal suffrage brought a new element into play. A chief might be of a different political persuasion, or belong to a different political party, from the government administration in power; or he might be sympathetic to the Government, but the generality of his people favour the other camp! Thus there was inherent conflict in the system. The Government might call on the chiefs to be above party politics, when it really meant that the chiefs should toe the party line! The local councils might have been elected to counteract the power of the chief (and in effect of Central Government). The bitter struggles between the various parties not uncommonly rendered it impossible to carry on participatory local government successfully. These problems may have been intensified in Nigeria by the creation of State governments. Expansion in the scope of participation inevitably causes an erosion of traditional authority. Perhaps one should return to a system where chiefs are dependent on a local income, so that they may be induced to become representatives of their people. It may be that in some localities the respect for and authority of the chiefs has decayed to a degree that is beyond resuscitation. The various reforms that have been tried have been half measures – attempting to reconcile conflicting impulses, both to accommodate democratic principles and to reinforce traditional values.

Adebayo writes of local government in western Nigeria, 'The party system is founded on the basic principle of ideological difference or complete divergence of social and political goals, [and] the citizens of a given political unit . . . rally behind the banner of whichever party satisfies their ideology and conscience. But here was a local community, constituted into a local government council. The functions and programmes of the council were distinctly enumerated by the instrument establishing the council; the manner of operation, procedure and the goals envisaged were spelt out in detail in the Local

Government Laws. All that was required of them was to raise sufficient revenue to execute the programmes and functions already specified clearly in the law and instruments. Where then is the place for the display of the foreign expensive luxury known as party politics?

'Yet this was precisely what occupied the mind and attention of the various councils most of the time. They dissipated their energy on the futile issue of which group constituted the "government" and which group the "opposition". In the process, a lot of heat and bitterness was generated, all to no purpose . . . local government soon ceased to be an institution for social and economic development at local level and became a battleground for politicians to test their strength. In the light of this, it seems clear that if local government institution is to succeed in the future, the basis of representation must be such as would bear no direct link or relation with national partisan politics. The main criteria should be to bring in persons who by their social standing and reputation in the community can effectively collect council revenue, honestly husband it and get on with the task of development at local level. It is, of course, not an easy task to expect the ballot box to come out with persons possessing these distinctions. Qualification for eligibility to vote may have to be fashioned in such a way that, as far as possible, it will exclude political fanatics and fortune seekers. Ballot in each ward of the council area should certainly not be on the basis of national party political labels.'[11]

At the start of this century, Ireland was still a part of the United Kingdom; Russell reviewed his experience as a rural reformer, and considered the development of rural cooperative movements *vis-à-vis* the establishment of responsible independent government. He wrote, 'The great problem before democracy is the evolution of a social order which will ensure, so far as anything human can be ensured, that the democracy will put forward its best thinkers, its wisest men of affairs, and that it will develop a respect for the man of special and expert knowledge. Unless the aristocracy of character and mind replace the demagogue, whose only talent is his fluency on a platform, and the skill with which he echoes back to an ignorant crowd the prejudices which populate . . . their skulls, our democratic systems of today will be no exception to the ancient law or cycle of political change formulated by Aristotle . . . Every kind of people get the politicians they deserve, and we must organise the nation so that people may be

11 Adebayo, A. (1981) *White man in black skin,* Spectrum Books, Ibadan (p. 104)

more deserving of, and more discerning of, better qualities in their public representatives than they are at present [in Ireland in 1912]. Unorganised individualism in a country where the small farmers who read little or nothing form a majority of the population will never lead to this knowledge. The promoters of the organisation of agriculture in Ireland are trying to create in every parish associations of men to help each other and to do their business together. These associations demand for their success men of scientific knowledge and business capacity at their head. As the prosperity of the individual in the association depends largely on the success of the communal enterprise, he rapidly develops public spirit and a desire for good leadership, and for the public welfare in which his own is implicated. These qualities in the parishes will become national attributes. They will permeate local government and national assemblies, and will bring about political stability and good government. These qualities can never be engendered by an unorganised rural community weltering in an anarchy of individual effort, unable, not having experience of them, to appreciate great qualities of mind or character or the value of special knowledge in affairs. The parish is the cradle of the nation . . .'[12]

The need for innovation and the role of owner-cultivation

In earlier chapters it has been shown that efficient and productive agriculture requires owner-cultivators who are innovative. It is realised that in the preliminary phase of developing commercial farming in Africa, the trend will be to the owner-manager – that is a farmer who directs labour rather than a farmer who does most of the work himself – but the objective must be owner-cultivation. A result of owner-cultivation is that farm size is restricted, because of the limited amount that one man can do, even if he is aided by machinery. Therefore the owner-cultivator requires the support of agricultural cooperatives and/or of marketing boards, which the farmers should be able to control in their own joint interests. The unfortunate fact is that, in Africa, Government has tried to run these institutions for its own benefit, and vitiated their usefulness. Can Government let illiterate or semi-literate farmers run them for themselves, when

12 Russell, G.W. (1912) *Co-operation and nationality,* Maunsell & Co, Dublin, reissued by the Irish Academic Press in 1982 (p. 103)

successful functioning depends on the comprehension and responsible behaviour of the members? Generally speaking the answer is that they could not. Though some cooperatives may work well because of the good sense and dedication of a key official, most would fall by the wayside. Nor can illiterate peasants be sufficiently innovative in their ways of farming. In other words, education is as vital for progress in farming as it is for progress in manufacturing industry. Galbraith writes, 'People are the common denominator of progress ... no improvement is possible with unimproved people ... Conquest of illiteracy comes first.'[13]

Initially, industry may require large numbers of human machine minders, using the mass production conveyor belt methods already well proved in the industrial countries; but the development of agriculture requires every farmer to combine scientific understanding and creative thinking. Progress in science depends on the use of literature, and we need to replace a 'verbal' society with a 'literary' society. This has happened already in Western Europe and North America. The priorities must be clearly perceived. In the forseeable future most people in Africa will only receive primary education, and this will be true particularly in rural areas. Therefore it is essential to make primary education as good as possible, and to ensure that the majority of people leave primary school with the basic skills of literacy and numeracy. To these should be added an understanding of their environment (both biological and economic), of science and of the techniques of innovation; they should also have a start in English, or whatever is the metropolitan language.

Creative behaviour may also be disruptive behaviour. The successful operation of cooperatives, and indeed of local government, depends on a society which is both disciplined and deeply imbued with a sense of personal responsibility. And for these to be instilled in the children, they must also be inculcated in the teachers. There is a tendency to copy teaching methods from developed countries, but these are often inappropriate, and not infrequently adopted just prior to their being abandoned as unsatisfactory in the former metropolitan country! Education is very vulnerable to the enthusiasms of ideologues. Fortunately, during my childhood, the soft options of civics, conservation, peace studies, etc, had not yet been discovered, but I am not aware of having suffered for the lack of them. Whatever is necessary can be incorporated in the traditional subjects of reading,

13 Galbraith, J.K. (1964) *Economic development,* Harvard University Press (p. 109)

geography, history, biology and so on. Primary education lays the foundations for all subsequent education, and the fundamental learning should be hammered home. As far as possible, it should be available to every person, without payment. Of course, nothing is really free, and in the end the community has to pay; this can be met out of community rates or taxes. Those who do not send their children to the community primary school will still have to pay their share. At the secondary school level the parents can contribute directly via fees, although it should be possible for exceptional children to win scholarships and to be awarded free places. Where the community is expected to maintain schools, then the control of the schools should reside with the community. Central government may provide financial subventions, but should content itself with setting and monitoring standards. It should have legislative as well as financial powers to enforce its control; this is already the situation in most countries. University education, on the other hand, is too costly for fees to meet more than an insignificant part of the expense, and should be, as far as possible, free. Entrance to university should be on a strictly competitive basis. The community as a whole is (via taxes) paying the bill, and therefore expenditure on higher education has to be related to the needs of society. Others who want the privilege of attending university must meet the full cost. It cannot be treated as the birthright of all who have the intelligence to benefit from it – or at least not yet in Africa.

The need for cooperation and the role of mortgage

Aspects of agricultural cooperatives have been discussed in Chapter 5, and are that:

- they enable independent cultivators to gear up the values of their lands, by means of mortgage, so that they may gain control of the activities ancillary to farming;
- these include purchasing, machine hire, breeding, processing and marketing;
- they give vertical integration of the agricultural industry, so that farmers benefit from profits at whatever level they occur;
- they protect the petty capitalist against big business, and to some extent against Government.
- marketing boards were originally forced cooperation of farmers, by means of legislation that enabled a majority of farmers to coerce an unwilling minority into collective action;

- cooperatives and marketing boards are primarily commercial associations, that are democratically controlled by the farmers in their own interests.
- success of the movement depends on educated farmers, who are motivated by a strong sense of public duty.

Farmers' unions, on the other hand, have a political function. Rather like trade unions in manufacturing industry, they represent the sectional interests of farmers in society. All three associations, that is cooperatives, marketing boards and farmers' unions, can serve as information-disseminating organs, advising farmers of new knowledge relevant to farming, of new techniques and of commercial opportunities. They can also offer a variety of services, such as advertising, banking, and insurance.

The need for government

We need enterprising farmers to solve the problems facing agriculture in Africa, and that is what private enterprise is all about! The ability of Government to innovate cannot compare with the manifold initiatives of many individuals. It is necessary to encourage businessmen to farm, but they will only accept to do so if the profits are comparable with what may be earned in commerce and industry, although land has the advantage that it may used as a security to borrow money, tends to keep or improve its value at times of inflation, and can be used to provide gearing for business activities. Not only will these men bring new ideas into farming, but their participation will help to restore status to farming. Moreover they have capital to invest in the development of appropriate farming systems, the independence to be innovative and to take risks, the conception of business management and control. It must be appreciated that the possible methods of farming are legion, depending on soil, climate and local market forces. But businessmen will wish to benefit from the enhanced land values arising from improved land use and increased profitability of farming.

Galbraith writes, 'The purposeful use of price incentives and disincentives can be a major substitute for administrative effort and time. No one doubts that good prices to farmers – and it is wholly consistent with effective use of the price system that these can be guaranteed by the State – are superior for increasing output to orders enforced by the police and magistrates. But likewise high import prices, the result of high tariffs and realistic exchange rates, may be more efficient than import licences for rationing scarce foreign exchange. It may be wise to exclude luxuries by fiat. But ability and willingness to pay may not be a bad test of the effective use of capital goods imports. Similarly, high interest rates, testing again the need of the community for investment, may be superior to a system of licences and permissions as a device for rationing capital ... The man who must pay the full price for imported goods has that much less incentive to offer bribes to obtain a licence. Even where the price system works less precisely than a system of licences, franchises, permissions and orders, there may be a strong case for it because of its remarkable capacity for eliminating the costs of delay.'[14]

Governments of developing countries may protest that their farmers devote themselves too much to export crops and too little to food production, but the peasant is only following his economic nose. If export crops pay better than food crops, he will cash crop for export rather than for food. An artificially maintained rate of exchange makes imported food cheap, and export crops relatively more valuable particularly where they can be exported outside the official market. Of course, if export crops pay better than imported food, then common sense suggests that it is better to export the crops and import the food.[15] However Government, and its battalions of civil servants and other appendages (what Cobbett called the 'Dead-Weight'!), would rather spend the foreign exchange so earned on buying aeroplanes, motor cars, ships, electronic equipment, foreign travel and other extravagances. Peasant farming is essentially private enterprise. If Government discourages export crops, it may end up without export earnings and without food! Moreover, there is a tendency to regard all food importation as a bad thing. In fact, it is only the ability to import food from regions outside their own that has banished periodic food shortages or famine due to inclement weather conditions from the developed nations. Witness the 'Hungry Forties' in Britain, when cereal crops and potato crops failed for successive wet years, and in 1846 the Government had to repeal the 'Corn Laws' that protected British agriculture, in order to allow the importation of corn from North America.

It is perhaps only since 1960 that the advantages of intercontinental trade, and the increasing rapidity of transport, have permitted the severity of famine to be mitigated in underdeveloped countries. It

14 Galbraith, J.K. (1964) *op. cit.*

15 I was informed by a World Bank consultant that the USA subsidises food production by about ten per cent and the EEC countries by about forty per cent, which cheapens their food exports but raises the cost of their manufactured goods to developing countries.

must be realised that, in times of necessity, countries require access to food supplies from outside their own weather systems, and they also need the resources to purchase them.

Countries whose main earnings of foreign exchange come from petroleum exports may argue that there is no advantage in devaluing their currency. This is true; but the artificially high rate of exchange is partly due to deficit spending ('Printing money') by Government, and partly due to neighbouring countries using a more internationally acceptable currency. Then Government must attempt to regulate the economy by means of import licences, price controls and criminal legislation, and by careful customs controls at its frontiers – rather than by giving its citizens freedom of choice and allowing the prices of goods and the value of the currency to find their own level. Government actions are arbitrary, and the effects are unpredictable. Even if economics were an exact science, the necessary data are sadly lacking in most developing countries. What is more, people will persist in acting in what they consider to be their best interests. Governments may pander to urban communities by favouring the importation of cheap food, which is assisted by an artificial rate of exchange. People in towns are concentrated, and thereby exert a greater threat to public order than the dispersed population in the countryside. A consequence is that rural areas become impoverished, and the drift to the towns is made worse. Eventually, a disaffected rural proletariat forms the grist for peasant guerrilla movements, and thereby for national disaster.

In developing countries, governments appear to wield much greater economic and political power than they do in industrial countries – perhaps than government has ever been able to do in industrial countries. This may be because government looms so large in the development process, which was formerly left to the whims and more or less unaided exertions of individuals. The smaller ruling establishment gives them an influence less trammelled by the inertia of the pervasive bureaucracy of the industrial nations. Today, there is much that only government can provide: currency, postal services, telecommunications, railways, roads, large-scale irrigation, etc. Sometimes this is because of the magnitude of the expense, or because of the degree of risk, or due to the sheer scale of the enterprise. Government has largely supplanted the traditional role in society of the chiefs; just as people looked formerly to the chiefs, now they look to government to solve their problems. Dumont writes that people 'have to be weaned away from the concept of the omnipotent State, in which individual contributions are meaningless'.[16]

16 Dumont, Rene (1966) *op. cit.*

Mrs Ellen Johnson-Sirleaf, former Minister of Finance in Liberia, has said, 'Economic development must rest largely on the judicious use and management of our scarce resources, [and] the first action on our economic agenda is to halt the corruptive haemorrhaging . . . Towards that end, government must reduce the level of its intervention in the economy, the level of bribery and kick-backs to a long line of civil servants. This can do no less than encourage private entrepreneurship to assume those activities which do not require government ownership or participation . . . The government intervention of which I speak is not merely in the economic sense . . . but in the form of personal harrassment most times for money, and official meddling in the affairs of private enterprise. Second on the economic agenda [should be] a reprioritising of government resource allocation. You may know that the biggest economic activity, the one accorded the biggest priority today, is the construction of public buildings – an activity which represents a non-productive investment. While agricultural and rural development programmes are on the verge of closure for lack of funding, a wide range of buildings – Ministry of Defence, Ministry of Health and scores of markets are being constructed . . . One of my market women friends put it in this way "what is the use of having a nice spot in a new building when you sit all day without selling $5 of goods because no one has the money or has taken pay to buy anything?" Our priorities must thus shift to productive endeavours in agriculture and agro-industries particularly those that are of small and medium size . . . Finally, the economic agenda calls for the restoration of confidence in the ability of Government to manage the affairs of state . . . Our government must be serious and, more importantly, must be perceived as serious in its work to reverse the economic slippage. Here it is a question of action and not words. Hence, no matter how passionate the plea for private investment and the call for a return of our lost human and capital resources, unless those making this request act in a manner consistent with this call, it is a waste of time.'[17]

Government has essentially the same problems as industry and agriculture – it has tried to copy the systems, and adopt all the impedimenta, of the industrial countries, without possessing the 'industrial ethic' of the industrial countries. This cannot be learned by sending a few people on overseas courses; it must come the hard way. In fact people do not sympathise with this 'European' way of going

17 Johnson-Sirleaf, Ellen speech to the Union of Liberian Associations in the Americas, Phiadelphia, July 1985, quoted *West Africa* No 3552, 23 September 1985

about life, for in terms of human relationships it is a retrograde step. Government bureaucracy does not operate particularly efficiently in industrial countries, but in developing countries it scarcely seems to function at all! Our virtues may often become our vices: compromising is temporising; initiative is opportunism; communicativeness is idle talk; communal solidarity is nepotism or tribalism; tolerance is reluctance to exert discipline or to punish. Government blames commerce for encouraging bribery, when this may be the only way to induce civil servants to do the work they are already paid to do.

Governments in developing countries tend to employ too many people and to reserve too few funds to permit anyone to do sufficient work. When recession comes, there is an inadequate margin beyond fixed costs to allow government expenses to contract. In the hope that the problem may go away of its own accord before too long, Government may delay paying the salary of civil servants, or more usually of teachers. 'Off the cuff' economies may prove to have alarming effects on the national economy. For example, strict control of import licences may cause whole industries to grind to a halt due to lack of essential spare parts or of raw materials, and worse still they may be forced into bankruptcy. If government mismanagement of the economy produces the final disaster, and the infrastructure collapses, people will have to return to scratching a subsistence from shifting cultivation, and to supply their other wants by traditional handicrafts. This may be hailed as a success, because it has made people return to the land. In fact it is an unmitigated disaster! As populations continue to grow, and the resource base – the soil – becomes increasingly over-exploited, there will be an ever worsening impoverishment of the nation.

The alternative is reform, but agrarian revolution is a discontinuity in economic progress. Before it, in a peasant agricultural economy, the normal fiscal incentives operate; after it in an emerging industrial economy they function again. Around the threshold of change, normal economic controls become unpredictable. In Nigeria food is expensive compared with in Europe and North America, but food production is still insufficient to feed the nation. The reason is that farmers' earnings are limited by the scale of peasant farming. Remunerative agriculture can only come with a change in the character of farming, and there is no easy progression between the two. The new systems of farming will produce greater profits from cheaper food!

As the country industrialises, people will learn the new relationships needed by an industrial society, and the efficiency of Government will also improve. Meanwhile the cure for the ailments of government is the same as for those of agriculture and industry. There should be robust and independent units of activity, and a strengthening of discipline, and thereby of trust. It may be unfashionable to mention it, but many of these features actually existed in anglophone Africa prior to 1960. Relatively small numbers of civil servants were able to administer very large numbers of people. It arose because of the circumstances of colonial government in the early years, and also perhaps because it was often initiated by soldiers, for whom bureaucracy was more of a chore than a way of life. The isolation and poor communications of a thinly spread administration meant that there had to be a high degree of delegation, discipline and trust; this continued into the mature phase of colonial government, and into the years immediately following the granting of national independence.[18] An important reason for its effectiveness was that the professional departments, including Agriculture, Education, Forestry, Geological Survey and Veterinary, were self-administering, with a professional officer at the head of each department. Virtually all senior administrative positions were held by professional officers. They controlled the appointments, promotions and discipline of both professional and clerical staff, and the latter did not usually make up more than ten per cent of the total staff. Compare this with the present situation in which the professional departments are headed by professional administrators, who have never worked in the field, and in which appointments, promotions and discipline of clerical and administrative staff are controlled not by the department, but by the Ministry or by a separate body altogether from either. This is the first step on the road to more bureaucracy and less work. For the generalised administrator, administration is an end in itself for which the exertion of power may be the only sublimation. This represents the power of veto, for he cannot do the work, and a few negative personalities can wreak havoc out of all proportion to their numbers. Undoubtedly some professionals seek power, but this may be mitigated by their reputations depending on professional achievement, and by working experience. In science and technology, information flow is essential. The professional forms a link in the chain of information flow, whereas the administrator tends to be a block in it. This may be due to ignorance, or to limited interest in

18 Colonial government is described by Aboyade as 'Minimum government'. See Aboyade, O. (1976) *Issues in the development of tropical Africa,* Ibadan University Press (p. 125)

professional matters, but it is not uncommonly because control of information is a facet of power and a way of becoming indispensable. These factors can operate at quite low levels within an organisation. The interpolation of general administrators into professional lines of command can sap an organisation of all immediacy. We need administrating professionals rather than professional administrators.

In fact this is more a problem of Britain than of other European or of developing countries, though as bureaucracy has grown it has increasingly affected other countries as well. In Nigeria, disregarding the irresponsible activities of some politicians, the efficiency of the civil service and public utilities continued to improve until the mid 1960s, when it was disrupted by coups and eventually by civil war. The final *coup de grâce* to government efficiency was given by the tremendous expansion of the civil services in the early 1970s. Nigerians, who had been catapulted into high positions at independence, may have forgotten that their own good performance was not due entirely to their own peculiar merits, but because they had taken charge of functioning organisations. Government was expanded rapidly, always by promoting from the bottom, because there was no other internal source of personnel. Inevitably, though, the point was eventually reached when inadequately experienced administrators were in charge of inadequately experienced subordinates. Nigerians who may be indignant at what I say may like to look at the 'Short form of service for Government offices' (overleaf) which is understood to have been composed and circulated through the British civil service some years ago! This reveals that the malaise is not confined to Africa; simply that it suffers from the same disease, but more severely.

The need for technical assistance

Mellanby, describing his experiences with the (then) Colonial Office, wrote, 'Notwithstanding the urgency stressed by all, not least those in the Colonial Office itself, it took some two months for a letter to appear. No-one is able to explain why these delays took place then, and still take place in similar circumstances today. One meets so many absolutely first-class people inside the Colonial Office, they show no lack of vigour and they understand the urgency of their problems, but for some reason the machinery of the office is such that delays much worse than mine are commonplace, and many posts have not been filled because the candidates got tired of waiting after having been told how important their work will be. Worst of all perhaps is the situation where a man has been kept waiting for months for a letter of

Short form of service for Government offices

Prayer Let us pray.
Oh Lord, grant that this day we come to no decision neither run into any kind of responsibility, but that all our doings may be ordered to establish new departments for ever and ever.

Aye men.

Psalm The Government is my shepherd; I need not work.
It allows me to lie down on good jobs. It leadeth me beside still factories. It destroyeth my initiative.
It leadeth me in the paths of the parasite for Politics' sake. Yea, tho' I walk through the valley of the shadow of laziness and deficit spending, I will fear no evil, for the Government is with me.
It prepareth an economic Utopia for me by appropriating the earnings of my grandchildren.
It filleth my head with bunkum and my inefficiency runneth over.
Surely the Government shall care for me all the days of my life, and I shall dwell in a fool's paradise for ever.

Hymn Oh Thou who see'st all things below,
Grant that thy servants may go slow;
That they may study to comply
With regulations till they die.
Teach us, Oh Lord, to reverence
Committee more than common sense;
Impress our minds to make us plan
But pass the baby when we can.
And when the Tempter seeks to give
Us feelings of initiative,
Or when, alone, we go to far,
Chastise us with a circular.
Mid war and tumult, fire and storm,
Strengthen us, we pray, with forms;
Then will Thy people ever be
A flock of perfect sheep to Thee.

Benediction The peace of Whitehall, which passeth all understanding, preserve your minds in lethargy, your bodies in inertia, and your souls in coma, now and for evermore.

Aye men.

appointment, so that he has been unable to make his own plans, and then he is asked to take up his duties in a few days.'[19] The inheritors of the Colonial Office are perhaps the technical aid agencies, the difference being that what took the Colonial Office months to achieve, the technical aid agencies can accomplish in an equivalent

19 Mellanby, K. (1958) *The birth of Nigeria's university,* Ibadan University Press

number of years! In the former era, the majority of people working in government service in the colonies (as they then were) were somewhat insulated from the excesses of the Colonial Office.

The technical aid agencies have exported the metropolitan bureaucracy to their outposts in the developing countries. It is possible that, behind the walls of embassy or high commission, these organisations can function more or less as the civil service does in the home country; but where it comes into contact with Africa, being that much more exigent, it can be even more paralytic than African bureaucracy.

In the 'corridors of power', economists and planners seem to be usurping the mantle that, in Britain, formerly cloaked those who had obtained their degrees in 'Greats' or in the Classical Tripos at the Universities of Oxford or Cambridge. Governments are being urged to impose ever more exigent systems of planning and control, presumably needing yet more economists and planners at the helm. However, my impression is that such people lack the 'gut feel' that comes from 'serving on the beat'. They resemble army commanders who have spent their whole service at GHQ, and have never been exposed to the field of battle.

I have witnessed senior government officers sitting round a table and deciding how the forest behaved and how it should be treated, and investing considerable sums of money for this purpose; but the forest would not, could not, did not respond as they thought. Later, they tried to perceive only what they wanted to believe. They were deaf to the cries of field officers, or replied that the reasons for failure were because their subordinates did not try hard enough! It is people, cattle, grass, trees, soil that ultimately determine what will succeed and what will not. This is why economists, planners, accountants go astray – they think in terms of the system rather than of the wilfulness of the constituents, unless in some abstract or statistical way. Information flow, between the centre and the periphery of an organisation *in both directions,* is essential for its vitality and responsiveness. The opinions of the troops may be blinkered by their immediate environment, but they cannot be ignored.

Perhaps one ought to explain the sorts of ways in which difficulties can occur. There is, say, a technical aid agreement between the two governments. Let us ignore the passage of years that may have occurred before this came to fruition; it is now actually installed. Each government has contracted to provide certain inputs. The host government may be eager to get the aid, either for the sake of the foreign exchange that it can bring to the country, or because it thinks the project may be of genuine benefit. It has therefore accepted conditions that it may be unable to fulfil, or subsequently may find itself unable to do so. Usually the donor country supplies the external inputs of technical expertise, technical equipment, and one hopes, the technical back-up that must be provided from overseas. The host country provides housing and heavy furnishing for the project staff, and office and working accommodation together with additional professional, technical and clerical personnel. It may also have agreed to service, repair and supply fuel to project vehicles; but perhaps the host institution has overspent its budget, or is in protracted hiatus at the beginning of the financial year, or the fuel supplier has not been paid owing to some bureaucratic bungling; in short, there is no petrol. Unless the technical aid experts can pay for fuel (or whatever) out of their own pockets, no transport will move and no work can be done. The problem will be referred to London/Paris/Rome/Washington, or wherever, and after a delay of several months the project may get permission to divert technical aid funds for this purpose, and work will begin again. The decision may have to be taken by some anonymous official in the labyrinthine bureaucracy of the metropolitan country. It takes two or three weeks for a letter to travel in each direction, the telex has not been functional for several months, and cables never arrive at all. Matters are sometimes made worse by the constant posting of aid administration staff, both at home and overseas, so that too few people seem to be familiar with what they are doing. Meanwhile the experts and their counterpart staff will have sat around for several months unable to go anywhere or to do anything. It costs about US$100 000 annually to maintain an expatriate in a developing country, without his doing a stroke of work! Moreover, valuable machinery and equipment may have been kept idle over this same period; or the planting or harvesting season may have been dissipating without vital operations being done. In a moment of despair, a French *cooperant*, who was supposed to be producing a manual for students, was heard to exclaim, 'If there is paper, there is no ink; if there is ink, there is no paper; and if there is paper and ink, the [duplicating] machine is broken down and there is no money to repair it!'

Once in exasperation, I wrote in a report to a technical aid agency that I had found more difficulty working for them than in fifteen years spent in the employ of an independent African country! Of course it is rash to write this kind of thing unless one's appointment is permanent and pensionable. The fact is that, in technical aid, virtually all the professional staff engaged in the field are on temporary contract employment, and only the clerical and administrative staff have permanent and pensionable terms of service. The few professional officers on pensionable service are called '*advisers*' so that too much notice need not be taken of them! A natural consequence of such arrangements is that technical aid is run mainly for the sake of the bureaucracy and not for the work. A few years ago, at a time of financial stringency, an aid organisation was requested to economise. It retrenched its field staff by 13 while at the same time increasing headquarters staff by 79! It is not unknown for the staff in the metropolitan headquarters to number about half of the total staff engaged on aid work. Neither bilateral aid nor international aid agencies are immune from these influences. It is of interest that, in recent years, some government aid funds have been disbursed through charitable and religious institutions – perhaps because they retain a cheaper and more devolved administration. Government is also turning to consultancies and management companies to assist in aid work, possibly to escape the impediments of civil service bureaucracy and the obligations to civil service staff, though aid managed in this way is not cheap. Aid institutions may blame African organisations for delays and lack of response, but they give exactly the same impression, but the other way round! Procedures that by long usage can seem like common sense, to other people may appear bewildering and nonsensical. Technical aid, if it is to succeed, must be run like an army in the field.

Technical aid personnel themselves have other difficulties. To quote a passage from a project report which put it graphically, 'The project adviser has to rely on the active good will of the local authorities concerned . . . Where such good will is forthcoming the programme can work . . . where it is lacking progress becomes impossible. The adviser comes to realise that he can only wave his arms in the wings, like an unwanted director at a scriptless performance where the actors would be happier to improvise without his interference.'

Difficulties may arise in various ways. There may be a defective relationship between the expert and the organisation in which he is working. A mutual rapport cannot be fashioned overnight, and may take several years to establish. Newcomers are treated warily, and at first may be regarded as a mild nuisance. The newcomer may not know how to get the best out of the host organisation in which he is lodged, and contract employees may face this same problem with the aid agency. Frequent promotions or transfers of staff in developing countries may result in the people who request the expert being different from those who eventually receive him, and who may resent his being foisted upon them. Counterparts (i.e. understudies furnished by the host institution) can have bad relations with the expert to whom they are assigned, or may leave for other employment before, or soon after, the project period ends. Nevertheless colleagues from the host institution, engaged on the same project, can be of considerable help in interpreting and manipulating their own organisation. However it is better if, as far as possible, working relationships resemble normal hierarchical lines of command.

The expatriate may be regarded as a status symbol, or as a source of reassurance, rather than being wanted to accomplish useful work. He can be used to legitimise courses of action already recognised as desirable, by stating the case or strengthening the arguments of the organisation concerned. On other occasions his ignorance of the true state of affairs may be exploited, and should the enterprise fail, he is conveniently at hand to absorb the blame. In this respect the contract employee is very vulnerable, and his career may be wrecked by matters which were in fact beyond his control.

People who set up projects, on both sides, are often administrators and/or diplomats, not unusually with limited understanding of the framework in which projects must operate. There is a tendency to leap to something new, rather than to persist with an established project. The relationships between aid administrators and field staff may be too remote and impersonal. Those who review projects may read all the various reports, but draw conclusions based on misapprehensions, which they do not discuss with those actually engaged on the field work. The project staff receive no overt response, so that they may wonder whether their reports are consigned unread to some bureaucratic limbo. Nevertheless, experts can be kept so busy writing project appraisals, budgets and work calendars, followed by periodic self-evaluations that, in the case of one expert who was supposed to report every month, he found it difficult to get time to do any useful work at all! One expert became so irritated that he began to write his reports in an increasingly startling style, and when this went unremarked, to write them as parodies of official documents. When this was noticed, his contract was not renewed and it is alleged that the aid agency attempted, by damning references, to ensure that he was never employed again. In fact he was an extremely popular officer, of more than usual effectiveness and ability. Usually a field inspection by a professionally competent officer quickly reveals the extent to which a project is achieving its objectives. Reports can be valuable in making people think about what they are doing, and the way they are doing it, but the production of reports in an elaborately prescribed manner is not infrequently to give a distant bureaucracy the sensation that it is 'controlling' events.

The aid expert is a temporary employee to whom no loyalty is owed, either by the giver or the receiver of aid. A consequence of this insecurity is that he can only condemn his project at the risk of losing his job. This may be because the project is abandoned, or because the last thing headquarters' staff want to hear is anybody's problems (despite their assertions), or worse still that their own performance should be criticised, however abominable it may be. The result is that the situation is whitewashed as much as possible, which also pleases the host institution. There is also another insidious effect: technical aid funds may be regarded as Government booty, the main aim being to scramble to get a share in it. If one's expertise is in some esoteric tropical science, there may be little choice. However the true purpose of aid, to hasten the development of Third World countries, becomes a secondary consideration.

A cause of protracted delays in implementing aid projects, is the need to obtain, from the host government, prior approval for each individual expatriate appointment. Doubtless this arrangement was originally to pander to the susceptibilities of newly sovereign territories. Today these may realise that we all have our hands in each others' pockets, and perhaps no country in the world can entirely escape the effects of this interdependence. If technical aid workers were permanent employees of institutions in the donor country, and these institutions were contracted to conduct the aid projects, then some of the delays and difficulties could be eliminated. The institution itself would be approved, on the basis of its established reputation, and time-wasting vetting of each individual member of a project would become superfluous. Of course, any member of a project who misbehaved could subsequently be declared *persona non grata,* but it is to be hoped that this would be a rare occurrence. With permanent employment, the institutions could select staff more carefully, and there could be administrating professionals rather than professional administrators. Overseas aid might then attract staff of a higher calibre, who could see in it some sort of career structure. Today, a person is foolhardy who chooses a profession of working in the field operations of overseas technical aid.

Improvement of technical assistance

Technical aid *can* be improved. I once worked in an institution where there were representatives from bilateral aid organisations of five different countries. I asked each how well they thought their project functioned, and how they thought this might be improved. Some clear threads emerged. Generalised aid agencies were not viewed with favour, because of their inherent over-bureaucratisation. Aid can be funded through parent organisations in the donor country comparable to those in which they are operating in the host country – for example university department to university department, or research institute to research institute. Staff can be seconded from the home institution and receive logistic support from it. Their conditions of service should be the same as for home-based staff, though they can receive extra privileges to compensate for the additional expense and inconvenience of working overseas. Mixed age ranges are essential – because organisations in developing countries tend to be deficient in older age groups. Young people find it easier to form social links across racial barriers, and older people with no previous experience overseas can find it difficult to adjust to unfamiliar circumstances. Staff working overseas need to be enterprising and resourceful.

Projects are often initiated by the aid-giving organisations rather than by the host country. The present situation is generally that aid-giving organisations think of something that ought to be done and then try to find the staff to do it; it would be better if an organisation acted the other way about, and tried to fit its objectives to the capabilities of its staff. In the developed nations there is a sufficient variety of institutions to accommodate most needs. Projects should have defined aims, approached by short term objectives which are reviewed and revised at intervals. Best results come from enduring and comfortable relationships between individuals of the host and parent organisations. Though individual projects may have a limited existence, the presence of aid personnel from the donor organisation may be continuous. Home-based and overseas staff would be interchangeable. The project leader, in the field, controls expenditure himself, which is sent to him in tranches, and accounted for in retrospect. This simplifies the administration, and avoids struggle with a disinterested bureaucracy whose comprehension of his needs may be limited. Then it becomes possible to attribute success or failure of a project to the individuals concerned. Dual financing by host and donor is not normally to be recommended. Past experience shows that this tends to disable a project, as it gives the leader problems of surmounting two different bureaucracies, with all the attendant difficulties and delays. An alternative may be re-imbursement of the donor by the host; but for various reasons it may not be easy to get the host government to honour its obligations. If a project is to function efficiently there cannot be a rigid interpretation of what costs should or should not be met by the donor. Moreover, a project needs insulating against infrastructural and executive weaknesses of the host country. In developing countries, budget control may be weak, so that funds may not be available when needed, and commercial organisations may refuse to honour purchase orders because of unsettled debts. Staff may be almost completely frustrated by the situation in which they find themselves. It could be supposed that if funds were made available to them, and the necessary support given, they would be released from this predicament. A Canadian organisation has attempted to finance projects in this way. If funds are given to institutions rather than to individuals, or inputs required from the host institution, then administrative weaknesses may cause money to be frittered away. Although this system may give useful lessons in the exertion of self-discipline and trust, it is apparent that the funds which could be disbursed, and the individuals and institutions that could be favoured, may be limited.

If aid administration is decentralised, then it is necessary to have a supreme committee to direct aid and allocate funds, in order to obtain an overall view of aid distribution both geographically and professionally. This would be supported by an inspectorate, both to monitor aid projects and to act as consultants to Government. If it misapplied its powers, the organisations through which aid is disbursed ought to be strong enough to appeal over its head. Administrative costs are reduced by using existing bureaucracies, and vulnerability to individuals by relying on a multiplicity of channels for aid. Generalised agencies may still be useful to support young volunteer workers, to service 'expert' personnel overseas and to act as an information and national centre; but projects should be permitted to run independently of the centre if they prefer it, or if they find that it does not support them adequately. As a local administrative agency, the centre can assist in pursuing proposed projects through the Ministries of the host country, which may otherwise founder in the bureaucratic toils or require repeated missions, sent at considerable expense from the donor country, to see them through. The centre can also identify suitable work for young volunteers, and deploy them. These are usually best placed where they are given organised work, usually of a technical nature, to do. It is not desirable if they are allowed to drift, or given organisational duties for which they are insufficiently experienced.

Aid programmes should not be doctrinaire, and the requirements for technical assistance differ widely between countries. Aid has advantages also for the donor country, in that it provides an understanding of the host country which assists in commercial or diplomatic decisions, and provides a mutual familiarity which should be of benefit to both host and donor countries and their citizens. However the balance between diplomatic/commercial/cultural/philanthropic interests may differ for each donor country and for each host country in which it operates. It may be argued that aid shores up indigenous inefficiency, and that, if it were not given at all, countries would learn to be self-reliant or to do without. However, where a donor country wishes to expand trade or invest substantially in another country, it has an interest in maintaining its infrastructure, and indirectly its political viability. The immigration of European or Asian skilled professionals into African countries is not regarded as desirable, and therefore alternative methods must be sought, even though they may be less efficient. Perhaps, in disbursing and controlling aid, there is an inherent attitude that it is a waste of money – and this may do much to ensure that it is often the case. Nevertheless, contributing to the development of Third World countries, and helping to sustain their internal organisation, may be equally important for sustaining and developing the economies of industrial countries.

Welfare in developing countries

Galbraith writes, 'Progress requires the radical elimination of retarding institutions. If elimination can be had from no other source, the Marxian alternative will sooner or later be tried . . . it is widely agreed that liberty and constitutional process are safer with the Western than with the Marxian alternative . . . The first task is . . . to build competent organs of administration. Then popular education. Finally social justice.' He also writes, 'If luxuries of the educational curriculum, esoteric educational institutions, refined agricultural services, prefabricated housing and a wide range of public services are adopted before their time, they will draw resources and energies from the tasks that are strategically vital for development. This is not beneficial, it is harmful. Let me press the point.' He continues, 'There is an assumption against which everyone responsible for an aid program must be on guard. It runs as follows: "The United States is a developed country; it has land-grant colleges, extension specialists, institutes for educational research and educational television. It follows that the country that seeks development must have these things." It does not follow at all.' He writes, 'Saving and capital accumulation are exceedingly painful in a poor country where the pressure of current need is very great. In a rich country savings may, of course, be excessive.' He continues, 'There are three things which the more advanced country has, which can be borrowed by those following it . . . (1) capital, (2) technology, (3) organisation.'[20] It is the last of these which may be the most difficult to transfer.

Allan writes, 'Private investment may be discouraged by limiting conditions based on admirable but unpractical concepts of social justice, in disregard of the Keynesian precept that "the engine which drives enterprise is not thrift but profit". The tendency in the developing countries is to want the best of both worlds – high wages, costly welfare programmes, national prestige, and rapid industrial and general economic development all at the same time. The unpalatable truth is that this cannot be done. Wages are a direct result of the investment of capital and they will not rise, nor will the number of jobs increase and national revenues expand, unless secure, substantial, and continuing profits are made by those who supply capital and enterprise. Social justice is one of the fruits of economic development and it is a characteristic of fruits that they cannot be reaped until the tree is mature. Patience is essential, but it is not a characteristic of the present temper of African nationalism.' He continues, 'When we consider the part that the land must play in the process of economic development we come back to the old dilemma. The economists point to the need for a crucial effort in the rural sector, but at the same time they agree, or most of them agree, that the development of this sector will be conditioned by progress in others, that economic growth must be sustained by industrialisation, and that a major increase of non-agricultural employment is essential. The conclusion is inescapable that *pari passu* with agricultural development must go industrial development, the two must be balanced and integrated, and that neither can proceed faster than the other . . . From the viewpoint of the agriculturist, the effects of industrialisation and general economic expansion are to take redundant men off the land, to create expanding markets for farm products, and eventually to leave the land to those who have the means and ability to farm it efficiently. Without this take-off, and if population pressure on the land continues to grow, the process of agricultural development will not only be retarded and limited, it will be reversed.'[21] As a Dutch rural sociologist said to me concerning 'African socialism', 'It is one thing to share out wealth, and another to share out poverty.'

Aboyade writes of Ceylon, 'The elaborate subsidies, the bad accounting practices and the discretionary management of exchange rates distorted both production costs and relative prices. The result was a perverse production system that misdirects the use of resources

20 Galbraith, J.K. (1964) *Economic development,* Harvard University Press (p. 109)
21 Allan, William (1965) *The African husbandman,* Oliver & Boyd (p. 505)

from the path of rationality. For example, given the low price of subsidised coconuts, the coconut peasant would rather collect his subsidised fertiliser and sell it to the higher-priced vegetable grower, refuse to grow coconut and then turn round to secure coconut products from the social welfare scheme . . . The network of subsidies had grown to consume about six per cent of the gross domestic product. With such an all-pervading welfare programme on a narrow resource base and weak economic structure, the entrenched subsidy-mindedness of the population shifted national preoccupation from production to consumption. The result was a growing rate of unemployment that has reached almost 15 per cent of the labour force, as high as 40 per cent among those aged fifteen to twenty-four years, and the astronomical level of 70 per cent among secondary school leavers [i.e. in 1967]. For a total population of only 13 million, there was an open unemployment of at least half a million . . . Yet, successive governments have been well-meaning in their desire for and commitment to social welfare . . . Administratively, it requires only social action to establish a welfare state. But for a poor underdeveloped country . . . it requires more than political will to sustain it. The ingredients of successful economic development and social change are always the same: hard-work, discipline, education and above all national management skills.'[22]

The duties of government

Government has four essential duties:

- maintaining a strong currency;
- law and order;
- national defence;
- communications.

It should try to carry them out supremely well. Where funds are insufficient, as many of the trimmings as necessary should be left to others to perform, though they may be encouraged by fiscal measures, and their performance monitored. The initiative and enterprise of Africans is at least as great as that of Europeans, and in the case of Nigerians probably greater. The worst excesses of private enterprise – the unacceptable face of capitalism – should be curbed by legislation, for example by imposing limited hours of work, a minimum wage, defined working conditions, minimum housing standards, and so on.

22 Aboyade, O. (1976) *Issues in the development of tropical Africa,* Ibadan University Press (p. 505)

However this should not be to the extent that either initiative is stifled, or the laws are ignored. Laws which attempt to control all facets of human life may have a stultifying effect on economic progress, because of the restrictions they impose on genuine enterprise. It is people who are the engine of development, not government, and excessive government intervention can do more harm than good – except perhaps in providing the legal framework in which change can occur. Undesirable developments, such as absentee landlords, rack renting, exploitive conditions for labour, can be dealt with when they become obnoxious. The inhumanity of man to man, the degraded lives to which so many were condemned, that was such a feature of early industrialisation in Europe, is something that one hopes will be minimised in Africa. Indeed many Europeans have really only emerged from poverty since the Second World War.

Nigeria, because of its size and relative wealth and the density of its population, could be the first country in tropical Africa to pass through the trauma of agrarian and industrial revolution. In preparation for this renaissance, the character of the people is beginning perceptibly to change. This has been particularly evident since the end of the civil war in 1970, and the start of the oil boom. The war gave pride and a sense of purpose to the nation, and people responded to the austerities and difficulties with self-reliance. This happened on both sides. I am not recommending the catharsis of war, but it did have these effects. Moreover the exodus of the majority of Ibos from the other regions of the country meant that others had to fill their niche, in particular the venturesomeness and enterprise, which had made them so successful in commerce and industry. In every age, there are peoples who culturally seem to come ready armed to meet new circumstances – the Portuguese in the 15th century, the Spanish in the 16th century, the Dutch in the 17th century, the British in the 18th and 19th centuries, and first the Americans and then the Japanese in the 20th century. It is the same within a nation. The rest of us benefit by copying the innovators. Where a section of mankind has forged ahead, the remainder soon try to follow. Whose turn will it be in the 21st century?

The significance of socialism

The change from peasant to industrial society cannot be achieved without abandoning a part of traditional culture. If we try to fit everything to the way that a traditional society functions, we shall never make progress. The reason for the success of mankind is that he has added cultural evolution to biological evolution; the former can proceed many times more quickly than the latter. It has enabled man increasingly to dominate the rest of the living world. The cooperative movement was initiated by Robert Owen. He has been called the 'father of socialism', and he happens to have been British. He lived at the time when Britain was going through the process that now affects Nigeria: the transition from rural peasantry to urban industrial society. In fact he was a factory manager, and he saw at first hand the misery, degradation and destruction of family life that was inflicted on ordinary people. He realised that society was heading towards a system, not of class interdependence, but of class conflict, and he foresaw that when the working classes had won (as they have more or less in Britain today) the habit of conflict would remain. Society would continue to be torn by the desire of each group to improve its position, no longer in relation to the masters, but in relation to other working class groups. He demonstrated by practical example the ability to bring the relationships of mutual dependence and reciprocal obligations of rural life into an industrial community. He tried to explain to others what he had achieved, and why he thought it had succeeded. His view is Utopian, and we cannot take all his ideas literally, for he was a pragmatist rather than a theoretician.

Alfred Wallace, who was the co-discoverer with Darwin of the theory of evolution, wrote concerning Owen's social philosophy, 'that the character of every individual is formed *for* and not *by* himself, first by heredity, which gives him his natural disposition with all its powers and tendencies, its good and bad qualities; and secondly by environment, including education and surroundings from earliest infancy, which always modifies the original character for better or for worse ... No one saw as clearly as Owen did how to put his views into practice; no one, perhaps, in private life has ever had such opportunities of carrying out his principles; no one has ever shown so much ingenuity, so much insight into character, so much organising power; and no one has produced such striking results in the face of enormous difficulties as he produced during the twenty-six years of his management of New Lanark ... Because he advocated the abolition of rewards and punishment, as both unjust and unnecessary ... it was argued ... such a system would lead to a pandemonium of vice and crime ... on the other hand Owen did ... transform a discontented, unhealthy, vicious and wholly antagonistic population of 2500 persons to an enthusiastically favourable, contented, happy healthy and comparatively moral community, without having recourse to any legal punishment whatever, and without, so far as appears, discharging any individual for robbery, idleness or neglect of duty; and all this was effected while increasing the efficiency of the whole manufacturing establishment ... [Formerly] the people lived almost without control in habits of vice, idleness, poverty, debt and destitution. Some were drunk for weeks together. Thieving was general and went on to a ruinous extent ... There was also a considerable drawback to the comfort of the people in the high price and bad quality of the commodities supplied in the village. The whole of the workers and overseers were strongly opposed to him ... though he soon introduced more order and regularity in the work ... An equally fierce opposition was made to any improvement in the condition of the houses and streets as to dirt, ventilation, drainage etc ... and for two years he made very little progress.

'One thing which he did for the benefit of the workers which had done some effect in disarming their enmity and suspicions. Instead of the retail shops where inferior goods were sold at credit for very high prices, he established stores and shops where every article was supplied at wholesale prices, adding only the cost of management ... But what at length satisfied them that their manager was their friend was his conduct when a temporary scarcity of cotton ... caused most of the mills to be shut, and reduced the workers to the greatest distress. But though Owen shut up the mills he continued to pay every worker for the four months during which the scarcity lasted, employing them in thoroughly cleaning the mills and machinery, repairing houses etc ... He was now at last able to carry out his plans for the education of the children, none of whom he would allow to enter the mills as workers till they were ten years old. He built handsome and roomy schools, playrooms and lecture rooms for infants from two to six, and for the older children from six to ten years old; and he obtained the best masters for the latter. The infant schools were superintended by himself, and managed by teachers he himself selected for their manifest love of children. His instructions to them were that they were on no account ever to beat any one of the children, or to threaten them in any manner in word or action, or to use abusive terms, but were always to speak to them with a pleasant countenance, and in a kind manner and tone of voice; that they should tell the infants and children that they must on all occasions do all they could to make their play fellows happy; and that the older ones, from five to six years of age should take special care of the younger ones, and should assist to teach them to make each other happy ... the children were to be taught the uses and nature or qualities of the common things round them, by familiar conversations when the children's curiosity was excited so as to induce them to ask questions respecting them ... The children were never kept at any one occupation or amusement till they were fatigued, and were taken much into the open air and into the surrounding country, where they were taught something about every natural object.

'Numerous visitors, latterly numbering two thousand every year, were more amazed and delighted with the schools than with any other part of the establishment . . . the dancing, the music, and all the other lessons and exercises out-of-doors, of the infants and children in their playground . . . their kindness of manner to each other, their unaffected joyous happiness . . . And truly those who have trained from infancy through these schools were by far the most attractive, and the best and happiest human beings I have ever seen. Their manner was unaffectedly graceful, and when spoken to by strangers, naturally polite, with great innocent simplicity. The total absence of all fear, and full confidence in and affection for their teachers, with the never-ceasing expression of a perfect happiness, gave these children of working cotton-spinners a character for their age superior to any I have yet seen. It was also noted how this training improved the physical appearance of the children, and many visitors declared that they had never seen so many beautiful girls and boys as in the schools at New Lanark.'[23]

It may seem that the book was intended to arrive at this theme, whereas I chanced to see these passages from Wallace as I was writing, and was immediately struck by their relevance. I must confess to knowing little about Marxism, which was a later social philosophy. Marxism has always appeared to be a political theory of class conflict, worked out in relation to industrial societies and then, by some mysterious illogicality, applied to the problems of peasant societies! Aboyade writes, 'The characteristic Marxian class division of society does not fit the emerging class structure in Tropical Africa . . . The Establishment is hardly yet deeply rooted as a socio-economic class.'[24] Robert Owen's reputation was somewhat dimmed when he tried to carry his ideas across to the United States of America. There he tried to create model communities, for example at 'New Harmony' in Indiana. The experiments failed. In the United States at that time, there was ample opportunity for all. Nobody wanted to work hard in order to support idle or incompetent members of society in equal comfort. In Europe it was then another matter – however clever, however competent, however hardworking a person was, he could not

23 Wallace, Alfred R. (1905) *My life – a record of events and opinions,* Chapman & Hall, 2 volumes (pp. 435 and 459)

24 Aboyade, O. (1976) *op. cit.*

easily improve his lot in life if he were born in the lowest stratum of society. People *had* to depend on each other in adversity; it was a question of 'United we stand, divided we fall!'

For most people in Britain, life has only become reasonably tolerable within the past quarter of a century. This has been mainly due to the growth of the national wealth, and also in some measure due to the creation of the 'Welfare State'. However, it may have come too late to have achieved all that it should. Vital as it is in an adolescent industrial society, it may be less essential in a mature wealthy society, when it may sap the self-reliance and sense of personal responsibility of its citizens – and this is something that was not expected. People need to feel that they are rewarded for merit, and if society is to progress, that they can control their own destinies. They do not value what is entirely free, and want to be able to strive for the benefit of their own children. The inducement is necessary, and the feeling is natural and universal. Consequently, in Britain today, we see the State Welfare Services becoming progressively more expensive and inefficient, while at the same time their quality deteriorates. State ownership of industry shows all the disadvantages of monopoly, with the additional disadvantage that commercial and political interests become confused. The Oxford Concise Dictionary defines socialism as 'the political and economic theory of social organisation which advocates that the community as a whole should own the means of production, distribution and exchange'. I do not think Owen would have been happy with the consequences, or even perhaps with the definition. In England we have recently witnessed the workers of two State-owned industries – the miners and the steel workers – almost literally doing battle with one another because of their conflicting interests. This is a far cry from Owen's vision that the objective should be to strive for national harmony and contentment.

Russell wrote, 'If capital wins we shall have the Servile State, and an immeasurable bureaucracy to keep the populace in order. If State Socialism wins humanity will have placed all its hopes on one system; and genius, temperament, passion, all the infinite variety of human life will be constrained by one policy.' Moreover, 'The sense of solidarity is not begotten by people all belonging to one large nation or great empire; but by union in local enterprises where a personal relationship can exist between the workers, and where success in labour promotes pride, and responsibility is provocative of thought, and experience in control generates wisdom, and we have the intensive cultivation of business ability and intelligence.' He also wrote, 'Working so, we create the conditions in which the spirit of the community grows strong. We create the true communal idea, which the Socialists miss in their dream of a vast amalgamation of whole nationalities in one great commercial undertaking. The true idea of the clan or commune or tribe is to have in it as many people as will give it strength and importance, and so few people that a personal tie may be established between them. Humanity has always grouped itself instinctively in this way.'[25]

Robert Owen demonstrated that the relationships, close, deferential and interdependent, which were thought peculiar to pastoral Britain, were capable of being reconstituted within industrial society. They could be reconciled with the machine and with the growing numbers of poor people. In fact his factory was one of the most productive and efficient of its time. The country which has, perhaps, most achieved Owen's ideal is, of course, Japan. One likes to think that their success owes something to his inspiration. The priorities of national development are to:

- generate national wealth as fast as possible;
- raise the majority of people to a tolerable quality of life as soon as possible;
- reward people for merit and hard work, and to harness personal initiative;
- prevent the exploitation of man by man, where some enrich themselves at the expense of others.

These all interact and a compromise has to be struck between them. Moreover capital has to be gathered together by individuals, by companies or by the State, or by combinations of the three. The danger comes when we deter the accumulation of wealth. We can easily replace a policy of national growth by one of national envy and mutual impoverishment.

As a postscript I should like to quote from John Bunyan's *Pilgrim's Progress* but to leave the last word with an elderly Polish refugee who recently sat opposite me on a train. He said, '*We human beings always make life so complicated for ourselves . . . we should allow evolution to take place, and then we should not have revolution.*'

25 Russell, G.W. (1912) *op. cit.*

The Slough of Despond

'This miry Slough is such a place as cannot be mended, it is the descent whither the scum and filth that attends conviction for sin doth continually run, and therefore it is called the Slough of Despond: for still as the sinner is awakened about his lost condition, there ariseth in his soul many fears, and doubts, and discouraging apprehensions, which all of them get together and settle in this place; and this is the reason of the badness of the ground.

'It is not the pleasure of the King that this place should remain so bad; his labourers also, have, by the direction of His Majesty's surveyors, been for above this sixteen hundred years, employed about this patch of ground, if perhaps it might have been mended; yea, millions of wholesome instructions, that have at all seasons been brought from all places of the King's dominions (and they that can tell, say they are of the best materials to make good ground of the place); if so be it might have been mended, but it is the Slough of Despond still, and so will be when they have done what they can.

'True, there are by the direction of the law-giver, certain good and substantial steps, placed even through the very midst of this Slough; but at such time as this place doth much spew out its filth, as it doth against change of weather, these steps are hardly seen, or if they be, men through the dizziness of their heads step besides; and then they are bemired to purpose notwithstanding the steps be there; but the ground is good when they are once got in at the Gate.'

The Pilgrim's Progress
John Bunyan
London 1678

Index of authors

Index of subjects